Powers of the Earth

Nate Needham

Copyright

Powers of the Earth

How the Energy of Plants Charges the Body and Awakens the Mind

Nate Needham

Contents

About the Author

Nate Needham is the founder of Fiboprana Publishing, a creative platform dedicated to exploring the limitless potential of human energy. For nearly fifteen years, Nate has lived as both an entrepreneur and an athlete, driven by a deep curiosity about what truly fuels the body and mind. His work bridges the wisdom of ancient traditions, such as Chi, Prana, and life force energy, with modern understandings of psychology, neuroscience, and human performance.

Nate's journey began with a personal search for balance. After years of extreme sports, business challenges, and inner struggle, he realized that true power doesn't come from pushing harder, but from learning how to align energy rather than force it. That insight became the foundation of Fiboprana's mission: to teach people how to cultivate their own life energy for clarity, strength, and transformation.

Through his books, Nate helps readers rediscover the body's natural intelligence and the subtle forces that connect all living things. His writing is practical yet profound, designed to make complex ideas about energy, consciousness, and vitality accessible to anyone seeking a better quality of life. Whether he's writing about reflection, breath, or the science of inner power, Nate's message is simple: the energy you cultivate within determines the world you experience without.

About Fiboprana Publishing

Fiboprana Publishing creates books and content that awaken human potential through the science and spirit of energy.

Our mission is to help people rediscover their natural vitality by understanding that life itself (sunlight, breath, thought, and movement) is powered by an intelligent, energetic design.

Each Fiboprana title blends modern scientific insight with ancient energetic wisdom, offering practical ways to elevate physical health, mental clarity, and spiritual connection. We explore how nature, consciousness, and the human body are part of one living system and how aligning with that system restores balance, strength, and creativity.

Founded on the principle that energy is the foundation of life, Fiboprana Publishing seeks to make this truth accessible to everyone. Through clear writing, timeless ideas, and a focus on transformation, we aim to inspire a global shift toward conscious living, one that respects both the measurable and the mystical aspects of being alive.

Fiboprana Publishing: Master Your Energy. Elevate Your Life.

Preface

Every human life runs on a current far older than we realize. Before there was electricity, before we learned to name "energy," the green world was already charging the planet leaf by leaf, pulse by pulse. Plants are not just the background of life; they are the living batteries that fuel every breath, heartbeat, and spark of human vitality.

Every inhale you take carries the charge of photosynthesis: sunlight transformed into oxygen, released into your bloodstream. Every meal is sunlight captured and condensed into matter. Every calm, grounded feeling you've ever had beneath a tree is more than psychological; it's energetic. Plants are the original transformers of light into life. And every human being is an extension of that system, a walking expression of the energy of Earth.

We've been told that plants give us oxygen, food, and medicine. What if they provide us with something more profound, a steady infusion of life force that nourishes both body and mind? Modern research and ancient wisdom are now meeting at this crossroads, showing that plants don't just sustain us, they charge us. The electric rhythms of trees, the ions in the air after rain, and the subtle frequencies emitted by leaves all interact with the human body's own biofield. We are not separate observers of nature; we are conductors in its living circuit.

When we spend time in green spaces, measurable things happen: our cells receive more oxygen, stress hormones drop, heart rhythms stabilize, and brain waves shift toward calm focus. Behind those numbers lies something even more extraordinary, an energetic exchange that recharges our system the way sunlight recharges a solar cell.

People who garden, walk barefoot on grass, or sit quietly among trees often describe the same experience: "I feel renewed. Grounded. Alive." Science calls it improved vagal tone, ion balance, or photosynthetic benefit. Ancient traditions call it *prana*, *qi*, or *life force*. Both describe the same truth: energy flows between all living things.

This book was written to illuminate that flow to explore the bridge between the measurable and the mysterious. In its pages, you'll find what science already knows about how plants power human life: how their oxygen fuels our cells, how their chemistry sustains our energy, and how their presence restores our nervous system. But you'll also step into the deeper side where intuition, tradition, and experience reveal that plants don't just feed the body; they feed the field that animates it.

Why this matters now is simple: our energy is fading. Modern life drains us faster than it replenishes us. We're surrounded by signals, screens, and artificial light but starved for the natural frequencies that align the human system. Meanwhile, the planet's living systems (forests, gardens, and wetlands) are being disrupted. The very sources of vitality that sustain humanity are being cut away just as we begin to understand how essential they are.

Yet there is hope, and it is green.

The power to recharge ourselves has never left us; it has been forgotten. Every forest, park, and houseplant is a living reminder that the Earth still wants to share its current with us. When we reconnect with breath, touch, attention, and care, our energy rises. We sleep better. We think clearly. We heal faster. We remember who we are.

By the end of this book, you will not only understand the science of nature's energy but also feel it moving through your life. You will see plants not as decorations, but as living partners in your vitality, allies in the art of staying charged, grounded, and alive.

Because when you tune into the energy of the green world, something remarkable happens: your body remembers how to make more energy than it spends.

Plants have always been here, quietly offering us the blueprint for limitless vitality.

The question is no longer whether they can recharge us; it's whether we are ready to receive.

Introduction

I didn't grow up surrounded by mountains or forests.

I grew up in the city, a place of concrete rhythms, glowing screens, and constant motion. Life there felt efficient, but rarely alive. I worked, I trained, I exercised, but the kind of energy I had was mechanical, the kind that fades when you stop moving.

Everything changed when I moved to the mountains. I went there to chase my love of extreme sports (snowboarding, mountain biking, and pushing limits). After moving there, something unexpected began to happen. As I spent more time outdoors, my energy didn't just increase; it transformed.

At first, I noticed it during my mountain runs. I used to time myself going up and down the trails, trying to beat my own record. But soon, the clock stopped mattering. I began to feel charged, not just stronger in my legs or lungs, but more awake in my mind. The more time I spent in nature, the less I depended on external fuel. I ate less. I drank less. Yet I felt more alive than ever.

It wasn't willpower or adrenaline. When I trained in the city or inside a gym, I could push my body, but I didn't feel that same renewal afterward. In nature, fatigue turned into flow. I could ride or hike for hours, even until dark, and somehow end the day more energized than when I started.

Over time, I realized that something subtle was happening, something science couldn't fully explain, but ancient wisdom already knew. The Earth itself was recharging me.

The more immersed I became, the more I felt connected to something infinite. My body needed fewer "supplies," yet I was running on more power. My mind was clearer. My emotions were steadier. My energy was abundant.

That realization changed everything.

I started studying ancient traditions, those that spoke of *prana*, *chi*, and the subtle life force that moves through all living things. I began reading scientific works on ecology, bioelectricity, and plant communication. The more I learned, the more the two worlds, ancient and modern, began to merge into one truth: energy is the language of life, and nature is the translator.

I came to understand that our relationship with the green world is not symbolic; it's electrical, chemical, emotional, and spiritual all at once. The air we breathe, the sunlight plants capture, the ions released from the soil, and the calm frequencies emitted by trees all of it interacts with the human energy field. When we're in balance with nature, our systems sync with hers. We generate more life than we consume.

I also discovered something more profound: the relationship is reciprocal. When I respect and protect nature, nature responds. When I move through the forest with awareness rather than haste, the air feels different and more alive. I've come to believe that the Earth listens, and when we move in harmony with it, it moves in harmony with us.

That's why I wrote this book.

Not to preach a lifestyle, but to share a truth I discovered through experience, that there is a never-ending source of energy and freedom available to every human being, and it costs nothing. It's the current that runs through the Earth, the plants, and every cell of our bodies.

My goal is to help you feel it too, to awaken that sense of connection so profoundly that you no longer think of energy as something you have to chase, but as something that's already flowing through you.

Because when you tune into the energy of nature, you don't just feel better, you come alive.

Chapter 1: The Oxygen Connection

The first time I stepped into the greenhouse, it felt like walking into a different world. Outside, the air was dry and heavy with summer heat, but inside, the space was alive. Tall tomato plants stood in neat rows, their leaves reaching upward like green hands stretching toward the light. The scent of basil floated in the air, mixing with the earthy smell of moist soil. Within minutes, I realized I was breathing deeper without thinking about it. The air felt crisp, almost sweet. I could feel my shoulders dropping, my body relaxing, as if my lungs had been given a quiet gift.

It was mid-morning, and the sun poured through the glass panels above, flooding the plants with light. I didn't know it at the time, but those plants were hard at work, turning carbon dioxide into oxygen with every second that passed. I stayed there for nearly an hour, writing notes and sipping water, feeling unusually alert and refreshed. It was easy to imagine I could spend all day in that room, just breathing in the quiet energy of the plants.

But later that week, I returned in the evening, long after the sun had dipped below the horizon. The greenhouse was the same place, yet somehow it felt different. The air was warmer, heavier. I took a deep breath and noticed it didn't have that same fresh snap it had in the morning. It wasn't unpleasant, but it was flatter, less invigorating. Without sunlight, the plants had switched from photosynthesis to respiration. They were now consuming oxygen and releasing carbon dioxide just as humans and animals do, waiting for daylight to begin the oxygen-making process again.

I didn't feel dizzy or short of breath, but there was a subtle shift. My head felt slightly foggy after a few minutes, and I found myself opening the greenhouse door to let in a bit of night air. It struck me then how completely I had taken oxygen for granted. Inside that sealed space, the difference between an oxygen-rich environment and one where oxygen levels dipped only slightly was noticeable. It was a reminder that the comfort and vitality I had felt in the morning weren't magic. It was biology, driven by the silent, constant labor of plants.

In that small greenhouse, I experienced in a few hours what our planet experiences every day on a massive scale. Sunlight fuels the green world, and the green world fuels us. When the balance tips, even slightly, we feel it. The difference is that on Earth, we can't just open the door to let more oxygen in. We rely on the forests, fields, and oceans to keep the air we breathe rich enough to sustain life. And that's a job only plants can do. The greenhouse story is the perfect place to begin, because what happened inside that glass-walled room mirrors a truth about life on Earth: the oxygen we breathe is a gift created every moment by plants. Without them, life as we know it could not exist. You may never have thought about where your oxygen comes from. Still, the answer is astonishing in its simplicity and in its fragility.

When I stepped into that morning greenhouse, the crispness of the air wasn't just a pleasant sensation. It was the direct result of one of the most important chemical processes on the planet: photosynthesis. Every breath you take, every move you make, is possible because this process has been running nonstop for billions of years.

The Green Energy of Life

Photosynthesis might sound like a complicated scientific term, but at its heart, it's a beautiful and elegant equation. Plants take in carbon dioxide from the air and water from the soil, and with the help of sunlight, they transform those raw materials into glucose, a simple sugar they use for energy, and oxygen, which they release into the air. It's like nature's version of a factory that runs on nothing more than light and water, producing both food for the plant and fresh air for us.

The magic begins with chlorophyll, the green pigment in plant leaves. Chlorophyll is like a solar panel at the microscopic level, absorbing energy from sunlight. That energy powers the transformation of carbon dioxide and water into glucose. The oxygen we breathe is a byproduct of this process, released when the water molecules are split apart. Plants don't "intend" to make oxygen for us, yet their survival strategy ends up keeping every oxygen-breathing creature alive.

Once oxygen is created, it has to leave the plant, and that's where tiny pores called stomata come in. Stomata are like miniature breathing holes, usually on the undersides of leaves, that open and close to let gases move in and out. During the day, when the sun is shining, stomata open to take in carbon dioxide and release oxygen. At night, the process shifts. Without sunlight, photosynthesis stops, and plants switch entirely to respiration, using oxygen and releasing carbon dioxide, just like we do. The greenhouse air I felt at night was proof of this subtle but important rhythm.

Not all plants contribute equally to our oxygen supply, and they don't all produce oxygen in the same environments. Towering rainforests, with their massive leaf surface areas, are some of the most efficient oxygen factories on land. Grasses, including crops like wheat and rice, also produce significant oxygen because of their global abundance and rapid growth cycles. Then there are the ocean's microscopic powerhouses: phytoplankton and algae. These tiny, floating plants produce at least half, possibly more, of the Earth's oxygen. Even though they're invisible to the naked eye, without them, the air you're breathing right now would not exist. On land, oxygen production also follows seasonal patterns. In temperate climates, it surges in spring and summer when plants are actively growing, and it slows in fall and winter.

Measuring the Invisible

You might be wondering how we can measure something as intangible as the planet's oxygen production. Scientists estimate that about 50 to 70 percent of our atmospheric oxygen comes from marine plants, with the rest coming from land-based vegetation. That balance is surprisingly stable over time, but it's not unshakable. Climate shifts, deforestation, and changes in ocean chemistry can alter those numbers.

Marine plants like phytoplankton work on an almost unimaginable scale. In a single day, the oxygen from these microscopic organisms can match that produced by all the rainforests combined. On land, tropical forests like the Amazon are critical not just for their oxygen output, but for their role in regulating global climate patterns.

When we shift the focus from global to personal, the numbers get more concrete. The average adult uses about 550 liters of pure oxygen every day. To support that, you would need several mature trees working around the clock. Multiply that by the eight billion people on the planet, and you begin to see how important it is that our oxygen-producing systems remain healthy and widespread. It's not as simple as planting a single tree per person. Different species, ages, and environmental conditions dramatically affect how much oxygen is produced, but the principle is clear: the more healthy plants and ecosystems we have, the safer our oxygen supply.

While Earth's total oxygen levels remain relatively stable thanks to long-term natural cycles, local declines can and do happen. In some polluted cities, oxygen levels are slightly lower than in cleaner rural areas, though the bigger problem is usually air quality, not oxygen depletion. In aquatic systems, however, oxygen loss is a serious issue. Runoff from agriculture can lead to algae blooms that eventually consume so much oxygen that fish and other marine life suffocate, creating "dead zones." On land, deforestation doesn't just reduce oxygen output; it disrupts the balance of carbon and oxygen in ways that can accelerate climate change.

Why Oxygen Matters Beyond Breathing

Oxygen isn't just a gas we inhale; it's the fuel that powers our cells. Every cell in your body uses oxygen to create energy in the form of ATP, the molecule that runs nearly every function of life. Without enough oxygen, cells can't produce energy efficiently, which is why low oxygen environments can leave you feeling weak, foggy, and tired. This is exactly what mountaineers feel at high altitudes: their bodies are struggling to keep energy production at normal levels.

Even a small drop in oxygen levels can affect cognitive performance. Think about a time when you've been in a stuffy room for too long and started to feel sluggish or had trouble focusing. That's your brain signaling that it isn't getting the oxygen it needs for peak performance. Hospitals often monitor oxygen saturation in patients because when oxygen delivery drops, the body's ability to heal is compromised. High oxygen availability supports immune function, speeds recovery from illness, and aids tissue repair after injury.

Oxygen levels are also a vital indicator of ecosystem health. In lakes, rivers, and oceans, scientists measure dissolved oxygen to determine if the water can support fish, shellfish, and other aquatic life. Low dissolved oxygen can signal pollution, over-fertilization, or temperature changes that threaten entire food chains. On land, the amount of oxygen a forest produces can act as an indirect measure of its overall health and balance. A forest under stress from drought, pests, or deforestation will often produce less oxygen and absorb less carbon.

When plant oxygen sources decline, the effects ripple outward. Less oxygen in aquatic systems means fewer fish, which affects bird populations, which in turn can affect seed dispersal for plants. On land, deforestation doesn't just impact the immediate area; it disrupts weather patterns, increases soil erosion, and can make surrounding regions less hospitable for both humans and wildlife. It's a chain reaction where one broken link weakens the entire system. Protecting oxygen-producing ecosystems isn't just about ensuring we have air to breathe; it's about maintaining the web of life that sustains everything else.

The oxygen you're breathing right now may have been produced days ago in a patch of seagrass in the Pacific, or this morning in a maple leaf halfway across your country. Every inhale is a connection to the plant world, whether you realize it or not. And every time we plant a tree, protect a wetland, or allow a forest to recover, we are not just saving plants, we are securing our own breath, our own energy, and the survival of the living systems we depend on.

Chapter 2: The Food Web Starts Here

When the small research boat approached the island, the scientists on board could already see that something was wrong. From a distance, the landscape looked bare and strangely colorless. Once lush with vegetation, the slopes now appeared dry and patchy, dotted with stunted shrubs and the skeletal remains of trees.

Years earlier, a handful of goats had been brought to the island, just a few animals, intended as a convenient food source for passing sailors. But goats are resourceful and voracious. With no natural predators and plenty of greenery to eat, their numbers exploded. They grazed relentlessly, stripping the island of grass, leaves, and even the bark of young trees.

At first, it seemed like a simple problem of lost plants. But as the scientists walked deeper inland, they saw the bigger picture. The birds that once filled the island with song were gone. Without plants producing seeds, fruits, or flowers, there was nothing to feed them. Insects had vanished too, their habitats destroyed. With fewer insects, the lizards and small mammals that relied on them had also disappeared.

The damage didn't stop at the shoreline. The absence of plant roots meant the soil was no longer held in place. Every heavy rain washed more earth into the sea, clouding the water and smothering the coral reefs that fringed the island. Fish populations declined. The vibrant, interconnected web of life stretching from the mountaintop to the ocean floor was unraveling before their eyes.

What struck the scientists most was how quickly it happened. In less than two decades, the removal of plants from the base of the food web had set off a chain reaction that crippled the entire ecosystem. Without plants, there was no foundation for life, no primary producers to capture sunlight and turn it into energy for everything else. It wasn't just a loss of beauty; it was a collapse of the system that fed, sheltered, and sustained every living thing on that island.

The lesson was unmistakable: when the green foundation crumbles, everything built on it goes with it.

When we think about food, we often picture the last step in the chain of meals on our plate. But the real story begins long before that, with a quiet miracle happening all around us: plants capturing sunlight and turning it into energy. Without this first step, no other step in the food web could exist. Every bite you take, whether it's an apple or a steak, traces back to a green leaf or an algae cell that turned sunlight into something alive that you could use. Plants are the starting point, the foundation upon which all other life builds.

Plants as Primary Producers

At the core of life on Earth is photosynthesis, the process by which plants, algae, and certain bacteria capture sunlight and store it in sugars. This sugar isn't just food for the plant; it's energy currency that can be passed along through the food chain.

Ecologists describe this in terms of "trophic levels." Plants and algae occupy the first level: primary producers. Herbivores, like rabbits, deer, or caterpillars, form the second level as primary consumers. Carnivores and omnivores, such as foxes, hawks, and humans, occupy higher levels. Without the first level, there's nothing to support the rest.

Imagine a pyramid. The broad base is plants. Every other layer, the insects, birds, mammals, even top predators, rests on that base. Remove it, and the pyramid collapses.

When you eat an apple, spinach salad, or bowl of rice, you're tapping directly into the energy plants have stored from sunlight. But when you eat a piece of chicken, you're getting energy that passed through the chicken from the plants it ate. At every transfer, some of that energy is lost to the animal for movement, warmth, and life functions.

This is why plant-based diets are more energy-efficient: they skip that extra step of energy loss. To produce a single kilogram of beef, for example, it can take up to 25 kilograms of feed and thousands of liters of water. A plant-based meal gets you closer to the source, using less land, water, and energy in the process.

Humans: the foundation of entire ecosystems. Leaves feed caterpillars, nectar feeds bees, seeds feed birds, and roots feed the soil's invisible armies of microbes. Pollinators depend on plants for their survival, and in return, they ensure plants can reproduce.

Even fallen leaves play a role, breaking down into the soil and feeding fungi, worms, and countless other small creatures. A single oak tree can host hundreds of insect species, dozens of bird species, and even mammals like squirrels, all connected through the food web the tree supports.

Nutrient Profiles of Plant-Based Foods
Plants provide the bulk of the world's carbohydrates, our most immediate source of energy, through grains like rice, wheat, and maize, and through tubers like potatoes and sweet potatoes.
They're also rich in plant-based proteins. Lentils, chickpeas, black beans, nuts, and seeds supply essential amino acids for building and repairing tissues. While some plant proteins are incomplete on their own, diverse plant diets easily cover the spectrum our bodies need.
And let's not forget healthy fats. Avocados, olives, chia seeds, and flaxseeds deliver omega-3 and omega-6 fatty acids, critical for brain health, hormone balance, and cell structure.
Micronutrients Unique to Plants
Beyond calories, plants are our primary source of vitamins like C, K, and folate, as well as minerals like potassium, magnesium, and calcium.
Spinach, for instance, packs magnesium, essential for muscle function and energy production. Bananas are famous for potassium, which helps regulate heartbeat and fluid balance.
Plants also produce phytochemicals, natural compounds like polyphenols and carotenoids, that protect our cells from damage. Blueberries, for example, are loaded with anthocyanins, which research links to improved memory and reduced inflammation.

Functional Foods and Superfoods
The term "superfood" gets thrown around loosely, but in scientific terms, it refers to nutrient-dense foods with proven health benefits. Turmeric contains curcumin, a compound with potent anti-inflammatory properties. Kale delivers high levels of vitamins, minerals, and antioxidants in a single serving.

A diet rich in a variety of plants helps lower the risk of chronic diseases like heart disease, diabetes, and certain cancers. Diversity is key: different plants offer different nutrient profiles, and no single food can do it all.

Role in Global Food Security

Three crops, rice, wheat, and maize, provide more than half of the world's caloric intake. Billions of people rely on these staples daily. In different parts of the world, regional crops like cassava, millet, and sorghum play the same role, adapted over centuries to local climates and cultural traditions.

This deep adaptation means plants aren't just feeding people, they're woven into identity and heritage. The maize fields of Central America, the rice paddies of Southeast Asia, the millet farms of Africa: these aren't just farms, they're the foundation of civilizations.

But this foundation is under pressure. Climate change brings heatwaves, unpredictable rainfall, and shifting growing seasons, threatening yields. Soil degradation, often from overuse of chemical fertilizers or unsustainable farming, reduces the land's ability to grow crops. Pests and plant diseases, some of which spread faster by global trade, can wipe out harvests.

These challenges make sustainable farming not a luxury but a necessity. Solutions are emerging. Advances in plant breeding and biotechnology produce crops that resist pests, require less water, or thrive in salty soils. Urban agriculture, rooftop gardens, hydroponic systems, and vertical farms bring food production closer to consumers, reducing transportation costs and emissions.

Diversifying our plant diets, moving beyond the handful of global staples, builds resilience. If one crop fails, others can step in. In a changing climate, diversity isn't just a culinary pleasure; it's survival.

Toolkit

Here are simple ways to put this chapter into action in your own life:

1. Eat Closer to the Sun: Choose one meal a day that is entirely plant-based. Focus on whole foods like vegetables, fruits, grains, and legumes rather than processed options.

2. Grow One Thing You Eat: Even if it's just herbs on a windowsill, growing your own food connects you directly to the plant at the start of the food web.
3. Try a "Plant Diversity Challenge": For one week, aim to eat at least 20 different plant species. This boosts nutrient variety and supports biodiversity.
4. Support Local Farmers: Shop at farmers' markets or join a community-supported agriculture program to strengthen local plant-based food systems.
5. Learn a Staple Crop's Story: Pick one staple crop (like rice or maize) and read about its history and cultural significance. Knowing its journey deepens respect for the plants feeding humanity.

Chapter 3: Natural Medicine Cabinets

In the cool, thin air of the Andes, the slopes were covered in forests unlike any seen in Europe. Jesuit missionaries, new to the region in the early 1600s, found themselves in an unfamiliar world not just of towering mountains and vivid skies, but of illnesses they had never encountered. One fever in particular was relentless. It burned through the body, left its victims shivering and weak, and, more often than not, ended in death. They watched as the local Quechua healers worked with quiet precision. When someone was stricken, they would trek into the high forest to a specific type of tree with smooth, reddish bark. Carefully, they scraped thin layers of it, dried them, and ground them into a powder. The powder was mixed into water or wine, and the patient drank it, grimacing at the bitterness. Within days, the fever would subside.

At first, the Jesuits were skeptical. European medicine of the time had no knowledge of such a cure. But desperation changes minds, and soon they, too, were seeking the bark. The word began to spread first through mission networks, then across oceans to Spain, Italy, and beyond. By the mid-17th century, the bark of this "fever tree," as it came to be known, was saving lives in malaria-ridden parts of Africa and Asia.

Centuries later, scientists isolated the active compound: quinine. For hundreds of years, quinine was the only effective treatment for malaria, a disease that had shaped human history as much as any war or empire. Without this Andean tree, the building of the Panama Canal might have been impossible. Explorers, settlers, and soldiers could not have survived in tropical zones where malaria was rampant.

The fever tree's story is more than a historical anecdote; it is a reminder that many of humanity's most significant medical breakthroughs began not in laboratories, but in forests, fields, and gardens. Long before plant-based medicines had chemical names, they were discovered, tested, and preserved by the people who lived closest to them. And sometimes, saving millions of lives starts with nothing more than a tree, some bark, and the knowledge to use it well.

The story of the fever tree is just one chapter in humanity's long and ongoing relationship with plants as healers. Long before white-coated scientists synthesized pills in sterile laboratories, our ancestors turned to roots, leaves, bark, and flowers for relief from pain, fevers, infections, and countless other ailments. Even now, in the age of advanced pharmaceuticals, plants remain the quiet, often overlooked partners in modern medicine.

Plant-Derived Pharmaceuticals

If you had lived five hundred years ago and fallen ill, your best hope for treatment would have come from the natural world. Willow bark tea for pain. Garlic poultices for infections. Poppy resin for pain relief was so strong it felt miraculous. For most of human history, plants weren't an "alternative" form of medicine; they were medicine.

What's remarkable is how much of that heritage still exists in our most trusted modern drugs. Many of today's pharmaceuticals are either directly extracted from plants or are synthetic versions of compounds that plants first created. Nature had the head start. Plants developed these chemical compounds not to heal humans, but to protect themselves against insects, fungi, competing plants, and even ultraviolet radiation. In doing so, they became living laboratories of chemistry, producing molecules that happen to work wonders in the human body.

Consider early remedies that have become pharmaceutical staples. The bitter bark of the willow tree, once chewed or brewed into tea to soothe aches, eventually gave us aspirin. The resin from the opium poppy, used in ancient Mesopotamia, is still the basis for some of the most effective painkillers we have today. In each case, human curiosity bridged traditional use and modern science.

Take aspirin. Its story begins with willow bark, which contains salicin. Ancient Egyptians and Native Americans both knew willow could ease pain and reduce fever. By the late 19th century, chemists refined salicin into acetylsalicylic acid, creating aspirin, a drug so common now that billions of tablets are consumed annually for everything from headaches to heart attack prevention.

Taxol, or paclitaxel, has a different kind of story. Found in the bark of the Pacific yew tree, it became one of the most powerful weapons against certain cancers, particularly ovarian and breast cancer. The compound works by interfering with the way cells divide, halting the growth of tumors. Without a tree growing quietly in the damp forests of the Pacific Northwest, thousands of patients might never have had access to this life-saving treatment.

The fever tree's bark, cinchona, gave us quinine, the first effective treatment for malaria. For centuries, it was a medical lifeline in tropical regions. Entire economies and global projects, like the Panama Canal, depended on quinine's ability to protect workers from the disease.

Then there's morphine, derived from the opium poppy. While its potential for addiction is well known, its role in pain management, especially in surgeries and for terminal illnesses, has made it one of the most important drugs in medical history. Without it, modern anesthesia and trauma care would be unrecognizable.

These are just the famous cases. Many other plant-derived compounds are quietly saving lives every day. The rosy periwinkle of Madagascar gave us drugs that treat childhood leukemia. Foxglove plants led to the development of digitalis, a life-saving heart medicine. Even marine plants, like certain algae, are being studied for antiviral and anticancer properties.

As of today, about a quarter of all prescription drugs are derived directly from plants. And many more are inspired by plant chemistry, even if they are synthesized in a lab. The pharmaceutical aisle, in other words, still owes much to the forest, the meadow, and the field.

Medicinal Herbs with Strong Clinical Evidence

Not every plant used in traditional medicine has withstood rigorous scientific testing, but several have. Echinacea, for instance, has been shown in some studies to support immune function and may slightly reduce the duration of colds. St. John's Wort has consistent evidence for improving mild-to-moderate depression, though it can interact dangerously with certain medications. Peppermint oil capsules are recognized as an effective treatment for some digestive issues, such as irritable bowel syndrome, with clinical trials supporting their use.

Other plants have gained validation through both centuries of use and modern science. Ginkgo biloba, with its fan-shaped leaves, has been studied for its ability to support cognitive function, particularly in older adults. Turmeric, a golden spice used in Ayurvedic medicine for thousands of years, contains curcumin, a compound with powerful anti-inflammatory and antioxidant effects. Garlic, beyond its role in cooking, can modestly lower blood pressure and cholesterol levels, contributing to cardiovascular health.

In each case, the knowledge didn't appear suddenly; it was passed down, often orally, through generations before science confirmed what people already knew from experience.

Even with proven herbs, caution matters. The dose can make the difference between a helpful remedy and a dangerous one. Purity and quality are also critical, since herbal supplements are not always as strictly regulated as pharmaceuticals. Some products may contain little of the active ingredient they claim or worse, contaminants. Choosing reputable brands, checking for third-party testing, and consulting healthcare professionals are important steps for anyone interested in herbal medicine.

Ethnobotany is the study of how people use plants, often focusing on traditional and Indigenous cultures. It is one of the most valuable tools in the search for new medicines. Many modern drugs like quinine, aspirin, and Taxol owe their discovery to the observations and practices of local communities who understood their environment intimately. Respecting this knowledge means recognizing it as intellectual property and ensuring that the benefits of new discoveries are shared fairly.

The path from traditional use to modern medicine is long. It starts with fieldwork, where researchers document how plants are used and collect samples. These samples are tested in the lab, often through bioassay-guided isolation, where scientists separate compounds and test them for biological activity. Promising candidates undergo further refinement, safety testing, and eventually clinical trials before they reach the pharmacy shelf.

The challenge is that many potential plant medicines may be lost before they're even discovered. Deforestation, climate change, and habitat loss threaten biodiversity and with it, the chemical treasures plants hold. Equally fragile is the traditional knowledge that guides scientists to these plants in the first place. Without cultural preservation and sustainable harvesting, we risk losing remedies that could save lives in the future. Plants have always been our medicine, whether we recognized them as such or not. The history of medicine is a story written in roots and bark, in leaves and seeds. Even in our modern world of synthetic drugs, every trip to the pharmacy carries the quiet legacy of forests and fields. And if we are wise, we will protect these living pharmacies for our health, and for the health of generations yet to come.

Toolkit
Here are simple ways to put this chapter into action in your own life:

1. Start a Personal Plant Medicine Journal: Choose three plants you read about in this chapter: one pharmaceutical origin, one proven herbal remedy, and one traditional remedy that interests you. Create a journal entry for each with the plant's name, what it's used for, how it works in the body, and any safety notes. Add sketches or printed photos to help you remember them. Over time, your journal becomes your own "mini pharmacopoeia" you can reference.
2. Try a Safe, Everyday Herbal Infusion: Pick a gentle, widely available herb with a strong safety record, like peppermint, chamomile, or ginger, and make a tea from the fresh or dried plant. Note the flavor, aroma, and any changes in how you feel afterward. This helps you tune into plants' subtler effects while staying safe.
3. Visit a Botanical Garden or Medicinal Plant Collection: Seek out a local botanical garden or university greenhouse that features medicinal plants. Seeing and smelling the living plants behind familiar medicines creates a deeper connection. Bring a notebook and jot down which ones surprise you most.
4. Learn the Safety Basics Before Using Herbs: If you want to experiment with herbal remedies, follow a simple checklist:
– Check for any medication interactions (consult a pharmacist or healthcare provider).
– Verify the correct plant species.
– Start with low doses.

– Use products from reputable, tested brands.

This not only protects your health but also builds good habits for any future herbal work.

Chapter 4: Plant Chemistry and the Human Body

The year was 1747, and the HMS Salisbury was cutting across the gray waters of the North Atlantic. The voyage had been long, too long. The ship's food stores, once fresh and abundant, had dwindled to salted meat, hardtack, and dried peas. Fresh fruits and vegetables were a distant memory. The men were growing restless, but worse than that, they were growing sick.

It began subtly. A sailor complained of unusual fatigue, his strength draining away even on calm days. Another had gums so tender they bled when he chewed. Bruises began appearing on the men's arms and legs without any injury. Soon, entire hammocks were filled with sailors too weak to stand. Teeth loosened. Old wounds that had healed months earlier reopened.

No one knew what was happening. Some blamed the "sea air," others muttered about curses. But among the crew was a ship's surgeon named James Lind. He had heard whispers that certain foods, such as cabbage, limes, and oranges, could heal sailors from this wasting sickness. On that cold spring day, Lind decided to run a quiet experiment.

He gathered twelve of the sickest men and divided them into groups of two. Each group received a different remedy: vinegar, seawater, cider, a medicinal paste, barley water, and to one lucky pair, two fresh oranges and a lemon each day. The results were astonishing. Within a week, the orange-and-lemon group was back on their feet, their gums firming, their energy returning. The others remained ill.

What no one knew at the time was that these sailors were suffering from scurvy, a disease caused by a lack of vitamin C, a simple molecule found in abundance in citrus fruits. Vitamin C is an antioxidant, protecting cells from damage and helping the body make collagen, which holds our tissues together. Without it, the body begins to quite literally fall apart.

This was one of history's first clear demonstrations of the power of a plant compound to save human lives. It wasn't magic. It wasn't superstition. It was chemistry, life-giving chemistry, forged in the leaves and fruit of a tree, carried across oceans, and delivered into the bodies of people who didn't yet know what a "vitamin" was.

Centuries later, the lesson remains the same: the chemistry of plants isn't just about flavor or fragrance, it's about survival. Every bite of fruit, every sip of herbal tea, every breath of scented forest air carries molecules that have shaped human health for as long as we've existed. The story of the HMS Salisbury and the miraculous recovery of two sailors from a handful of citrus fruit reminds us of a simple truth: plants are not just passive greenery in the background of our lives. They are living laboratories, producing a dizzying array of chemical compounds that can heal, energize, and protect us sometimes in ways we are only beginning to understand. For most of human history, people didn't know the precise molecular names of these compounds. They just knew that certain leaves soothed pain, certain fruits restored strength, and certain aromas lifted the spirit. Today, science allows us to peer into the molecular world of plants and see why they have always been our closest allies in health.

The Language of Plant Compounds

Antioxidants: Cellular Protectors

Inside every cell of your body, countless chemical reactions keep you alive, breaking down nutrients, generating energy, and repairing damage. But these reactions also produce unstable molecules called free radicals. Left unchecked, free radicals damage DNA, proteins, and cell membranes, accelerating aging and increasing the risk of chronic diseases. This is where antioxidants come in. They are like molecular peacekeepers, donating an electron to neutralize a free radical without becoming unstable themselves.

Plants are rich in antioxidants because they need them to survive their own exposure to sunlight, oxygen, and environmental stress. Vitamin C, abundant in citrus fruits, berries, and peppers, is one of the most famous examples. Vitamin E, found in nuts and seeds, protects cell membranes from oxidative damage. People who regularly eat antioxidant-rich foods tend to have lower rates of heart disease, certain cancers, and cognitive decline, not because antioxidants are magic pills, but because they reduce the invisible, daily wear and tear on our bodies.

Polyphenols and Flavonoids

Polyphenols are a vast family of plant compounds with potent antioxidant and anti-inflammatory properties. Flavonoids, a subgroup of polyphenols, give plants their vibrant colors and protect them from UV radiation and pests. When we consume these compounds, we inherit their protective benefits.

Green tea, for example, is loaded with catechins, flavonoids shown to improve cardiovascular health by lowering blood pressure and reducing LDL cholesterol oxidation. Cocoa is rich in flavanols, which improve blood flow and may enhance brain function. Berries provide anthocyanins, which help protect neurons from oxidative stress, potentially slowing cognitive decline as we age. What's striking is that these compounds often work in synergy with other nutrients in the food, producing a health effect greater than any isolated supplement could deliver.

Terpenes and Aromatic Chemistry

If you've ever inhaled the sharp scent of a pine forest, peeled an orange, or crushed fresh rosemary between your fingers, you've experienced terpenes. These aromatic compounds are part of a plant's chemical defense system, warding off pests and attracting pollinators. Limonene, found in citrus peels, has been studied for its mood-elevating and antimicrobial effects. Pinene, abundant in pine needles and rosemary, may improve alertness and support respiratory health. In herbal medicine and aromatherapy, terpenes are used not only for their scents but also for their biological effects, which can include calming the nervous system or supporting immune function.

Fighting Inflammation and Disease

Inflammation as a Root Cause

Inflammation is the body's natural response to injury or infection, a short-term, protective process. But when inflammation becomes chronic, it stops being helpful and starts fueling disease. Chronic inflammation has been linked to heart disease, cancer, Alzheimer's, and many autoimmune conditions. Plant compounds play a key role in dialing down this persistent immune overdrive.

Curcumin, the bright yellow pigment in turmeric, can block inflammatory pathways in the body. Resveratrol, found in grapes and berries, has anti-inflammatory effects on blood vessels. Even everyday foods like olive oil contain anti-inflammatory molecules such as oleocanthal, which works similarly to ibuprofen at a chemical level.

Immune System Modulation

Some plant compounds don't just fight inflammation; they help fine-tune our immune system. Beta-glucans from oats and certain mushrooms prime immune cells to respond more effectively to threats. Garlic contains allicin, which has antiviral and antibacterial properties. Spices like cinnamon and cloves have compounds that slow the growth of harmful bacteria.

Diet diversity matters here. Eating a wide range of plant foods exposes the immune system to different compounds, training it to recognize and respond to a variety of challenges while also reducing the risk of overreaction, which is a hallmark of autoimmune disorders.

Preventive Nutrition

Large-scale population studies consistently find that people who eat plant-rich diets live longer, healthier lives. This isn't because of a single "miracle nutrient" but because of the combined, synergistic effects of hundreds of compounds working together in whole foods. In many cases, whole plants outperform isolated supplements because the nutrients are delivered in their natural matrix, where they interact with fiber, enzymes, and other compounds in ways we don't yet fully understand.

How the Body Absorbs and Uses Plant Chemistry

Bioavailability Basics

Eating a nutrient-rich food is one thing; absorbing and using those nutrients is another. Fat-soluble compounds like vitamin E or certain carotenoids require dietary fat for proper absorption, which is why drizzling olive oil over your salad can help your body access more antioxidants from the vegetables. Cooking can break down cell walls, making compounds like lycopene in tomatoes more available, while in other cases, heat can degrade delicate compounds like vitamin C.

Your gut microbiota (trillions of microbes living in your intestines) also plays a massive role in processing plant compounds. Some polyphenols, for example, are only transformed into their active, health-boosting forms after gut bacteria break them down.

Metabolism and Distribution

Once absorbed, plant compounds travel through the bloodstream, sometimes targeting specific tissues. Flavonoids might accumulate in the brain, terpenes might concentrate in the lungs, and carotenoids may settle in the skin, where they help protect against UV damage. Metabolism varies between individuals depending on genetics, gut health, and overall diet, which is why the same food can affect two people differently.

Limits and Cautions

Just because plant compounds are natural doesn't mean more is always better. Some can be toxic in high doses, such as certain alkaloids or fat-soluble vitamins. Others can interact with medications; for example, grapefruit contains compounds that interfere with enzymes responsible for drug metabolism. A balanced, diverse diet remains the safest and most effective way to benefit from plant chemistry, rather than relying on excessive supplementation.

The chemistry of plants is, in many ways, the chemistry of life itself. From the vitamin C that saved sailors on long voyages to the polyphenols that protect our hearts, these compounds form an invisible bridge between the plant world and our health. They are reminders that every meal is an opportunity to nourish not just our hunger but the very processes that keep us alive.

Toolkit

Here are simple ways to put this chapter into action in your own life:

1. Antioxidant Power Plan
• Eat the rainbow daily. Aim for at least five different colors of fruits and vegetables each day to cover a wide spectrum of antioxidants.

• Include vitamin C sources (citrus, kiwi, peppers) and vitamin E sources (nuts, seeds, avocado) regularly.

• Swap one processed snack per day for an antioxidant-rich choice like berries, cherry tomatoes, or green tea.

2. Polyphenol & Flavonoid Boost

• Drink 1–2 cups of green or black tea daily for heart-protective flavonoids.

• Incorporate a small serving of dark chocolate (70% cocoa or higher) a few times a week.

• Add berries to your breakfast and rotate types (blueberries, blackberries, strawberries) to maximize variety.

3. Terpene Therapy

• Use fresh herbs like rosemary, thyme, and mint in cooking to increase terpene intake.

• Peel citrus fruits yourself to release mood-lifting limonene from the zest.

• Take a mindful "forest walk" once a week to inhale natural plant terpenes (known as forest bathing).

4. Anti-Inflammatory Eating Habits

• Add turmeric to soups, stir-fries, or smoothies; pair with black pepper for better curcumin absorption.

• Replace some animal fats with extra virgin olive oil for its anti-inflammatory oleocanthal.

• Include garlic in your meals several times a week to harness its immune-boosting allicin.

5. Immune System Resilience

• Eat at least 30 different plant foods per week (including fruits, vegetables, herbs, nuts, seeds, legumes) to support gut microbiota diversity.

• Add oats and mushrooms to your diet for beta-glucans that help prime immune cells.

• Rotate herbs and spices like cinnamon, ginger, cloves, basil to introduce varied antimicrobial compounds.

6. Maximizing Bioavailability

• Pair fat-soluble nutrients (carotenoids in carrots, vitamin E in spinach) with healthy fats like avocado or nuts.

• Lightly steam cruciferous vegetables (broccoli, kale) to retain vitamin C while making certain antioxidants more available.

• Eat a mix of raw and cooked produce for optimal nutrient variety.

7. Safety and Balance

• If taking medications, check for interactions with certain plant foods (e.g., grapefruit and some prescriptions).

• Avoid excessive supplementation unless advised by a healthcare professional—whole foods remain the safest source of plant compounds.

• Keep variety as your guiding principle: no single plant food is a cure-all.

Quick Start Challenge:

For the next 7 days, log the plant foods you eat. Try to hit 30 different types by the end of the week. Notice how your energy, digestion, and mood respond.

Chapter 5: The Healing Power of Green Spaces

In the heart of Medellín, Colombia, summers were becoming unbearable. The air shimmered with heat, traffic fumes lingered between concrete walls, and daily life slowed to a crawl. Residents said it felt like the city was "holding its breath" during the hottest months. Shops closed early, older people stayed inside, and children avoided playing outdoors. Health officials warned of rising heat-related illnesses, while meteorologists predicted the city's climate would only grow more extreme in the coming decades.

Faced with this challenge, Medellín's leaders decided to try something bold, something both radical and straightforward. Instead of investing only in more air conditioning or building higher walls of concrete, they turned to plants. The city launched what would become known as the "Green Corridors" project. Over the next several years, workers and volunteers planted thousands of trees, shrubs, and climbing vines along the busiest streets, riverbanks, and pedestrian paths.

The change was not instant, but it was remarkable. As the greenery grew, the once-harsh sunlight softened under leafy canopies. The branches and leaves didn't just provide shade; they actively cooled the air through a process called evapotranspiration, where plants release moisture into the atmosphere. Temperatures in these corridors dropped by up to 3 degrees Celsius, and air quality noticeably improved. The corridors also gave birds, butterflies, and pollinators new habitats, turning noisy, traffic-filled routes into vibrant, living spaces.

People began walking and cycling more often, shopkeepers stayed open longer, and neighborhoods once avoided for their heat and pollution became lively community spaces again. What started as a climate adaptation plan became a movement for urban renewal. Medellín's Green Corridors proved that plants were not just decoration; they were allies in stabilizing the climate, protecting public health, and improving quality of life.

When Medellín, Colombia, planted its "Green Corridors," the transformation was more than just a visual upgrade; it was a reminder of something we often forget: humans are designed to live in connection with nature. In the shade of trees, surrounded by the rustle of leaves and the subtle aroma of flowers, our bodies and minds shift into a state that concrete and steel simply can't replicate. This chapter explores how direct interaction with plants and natural environments isn't just pleasant, it's deeply medicinal.

Forest Bathing and the Immune System

In the early 1980s, Japan was facing a problem familiar to much of the modern world: rising stress levels, burnout, and an increase in stress-related illnesses. In response, the Japanese Ministry of Agriculture, Forestry and Fisheries introduced a term that would later spark a global wellness movement: Shinrin-yoku, or "forest bathing."

Despite the name, it doesn't involve water. Forest bathing is the practice of mindfully immersing yourself in a natural, forested environment. You walk slowly, notice the patterns of light through the leaves, breathe deeply, and let your senses guide you. It's not a hike to reach a summit or a workout for calorie burning; it's a deliberate slowing down.

The difference between a regular walk in the park and Shinrin-yoku is the focus. In forest bathing, you're not rushing to get somewhere. You're absorbing the atmosphere, tuning in to subtle sounds like birdsong or the crunch of twigs underfoot, and smelling the natural scents carried on the breeze.

Japanese researchers began to notice something extraordinary while studying forest bathers. Participants who spent two or three days in forest environments showed a marked increase in natural killer (NK) cell activity. NK cells are a type of white blood cell that patrols the body, hunting down virus-infected and cancerous cells.

The key seems to be phytoncides, the aromatic compounds released by trees as a defense mechanism against pests and pathogens. When we inhale them, these compounds appear to trigger a positive immune response in humans. Some studies have shown these NK cell benefits can last for up to a week after just a few hours of forest exposure.

Imagine that simply walking in the woods can tune up your immune system in ways that persist long after you've left. It's nature's version of a slow-release supplement.

Since the 1980s, forest bathing has moved far beyond Japan. Countries like South Korea, Finland, and the United States have established official forest therapy programs. In Europe, "green prescriptions" are becoming more common, where doctors advise patients to spend structured time in natural environments. Wellness retreats now often include forest bathing as a core activity, and even some corporate wellness programs have integrated it into employee stress-reduction plans.

While cultures may adapt the practice differently, using pine forests in Scandinavia, rainforests in Costa Rica, or coastal trails in California, the underlying principle remains the same: intentional, sensory immersion in nature changes the way our bodies function.

Nature's Role in Stress Reduction

Stress is a survival mechanism, but in modern life, it often operates in overdrive. Cortisol, the body's primary stress hormone, is meant to spike in moments of danger and then return to baseline. But chronic stress keeps it elevated, leading to immune suppression, higher blood pressure, and increased risk of disease.

Studies comparing cortisol levels before and after time spent in green spaces show a consistent trend: spending time in nature significantly lowers cortisol levels. In fact, a 20-minute walk in a park can produce measurable reductions. Interestingly, the drop is greater when compared to walking in busy urban environments, even if the total physical activity is the same.

Nature does more than calm our bodies; it resets our minds. Psychologists call this Attention Restoration Theory. The idea is that our brains have two types of attention: directed attention, which we use for tasks, and soft fascination, which is engaged when we look at a sunset or watch leaves move in the wind.

Urban life constantly drains our directed attention. Nature, with its gentle patterns and soothing unpredictability, gives it a chance to recover. That's why after a short break in a garden, we often return to our work with clearer thinking and renewed focus.

The sensory details of nature, like the smell of fresh rain, the feel of bark under your fingers, or the layered sound of birds and insects, create a kind of mental reset button.

Time in green spaces has been linked to reduced symptoms of anxiety and depression. But beyond individual health, nature fosters community. Parks, gardens, and tree-lined plazas encourage social interaction, from casual conversations to organized gatherings.

Community gardens are a great example. They not only provide fresh produce but also create a shared sense of purpose and cooperation. When people come together in green spaces, they often feel more connected to each other and to the place they live.

Plants in Healthcare Settings

Hospitals are increasingly integrating gardens into their designs not as decorative extras, but as active components of patient recovery. Studies have shown that patients with a view of trees from their hospital bed recover faster and require less pain medication than those who face a brick wall.

The famous example is from Roger Ulrich's 1984 study, which compared gallbladder surgery patients with different window views. Those with tree views left the hospital nearly a day sooner on average.

Even a few potted plants in a hospital room can make a difference. Indoor plants improve air quality by filtering out certain pollutants, but their benefits are not purely physical. Patients often report feeling calmer and more positive when surrounded by greenery.

Flowers, in particular, seem to have an immediate emotional impact. A vase of fresh blooms can brighten not only the room but also the patient's outlook.

Healthcare architects now face an exciting challenge: designing spaces that are both sterile enough for safety and green enough for healing. This includes careful plant selection, safe soil containment, and layouts that make greenery visible from as many patient rooms as possible.

Some hospitals use native plants to connect patients to the local landscape, while others choose aromatic species to subtly enhance mood. In all cases, the goal is the same: use the natural world to accelerate recovery and improve quality of life.

Toolkit

Here are simple ways to put this chapter into action in your own life:

1. Try a Mini Forest Bathing Session

You don't need to travel to Japan to experience Shinrin-yoku. Find a nearby park, nature reserve, or tree-lined path. Leave your phone in your pocket, walk slowly, breathe deeply, and notice the scents, colors, and textures around you. Even 20 minutes can shift your mood and physiology.

2. Schedule "Green Time" Like an Appointment

Just as you'd book a doctor's visit or meeting, put regular nature time on your calendar. This could be a morning walk in a park, lunch in a garden, or weekend hikes. Treat it as non-negotiable maintenance for your mental and physical health.

3. Pair Movement with Nature Exposure

If your schedule is tight, combine exercise with green time. Jog through a park instead of on a treadmill, do yoga in your backyard, or take work calls while walking outdoors.

4. Create Micro-Green Spaces Indoors

Even a small apartment can host plant life. Choose easy-care plants like pothos, snake plants, or peace lilies. Place them where you spend the most time, by your desk, near your bed, or in the kitchen.

5. Use Nature as a Stress Reset

When you feel tension building, take a "green break." Step outside, focus on the horizon, notice the movement of leaves, or listen to birds. This brief sensory shift can lower cortisol and help you return to tasks with more focus.

6. Volunteer in a Green Project

Community gardens, tree planting initiatives, and park cleanups not only give you nature time but also connect you with others. The combination of social connection and outdoor activity amplifies the benefits.

7. Green Your Workspace

If you work indoors, bring nature in. Add plants to your desk, place your workspace near a window, or use nature-inspired art and screensavers. If possible, arrange meetings in a nearby park or rooftop garden.

8. Seek Out Healing Gardens in Your Area

Hospitals, wellness centers, and botanical gardens often have spaces designed for restoration. Make a point to visit regularly, even if you're not there for medical reasons.

9. Experiment with Plant Aromas

Bring in the phytoncide effect at home with essential oils from trees like cedar, pine, or cypress. Diffuse them while you rest or meditate to mimic the calming, immune-supporting properties of forest air.

10. Track Your Nature Benefits

Keep a short journal of your mood, stress levels, or sleep quality before and after regular green space visits. Over time, you'll see patterns that reinforce why it's worth making nature a priority.

Chapter 6: Nature and Mental Health

When the old textile factory closed down, it left behind more than just job losses; it left a yawning, empty lot in the middle of the neighborhood. Over the years, weeds claimed it, graffiti covered the brick walls, and the only visitors were stray cats and the occasional trespasser. For the people living in the surrounding apartment blocks, it was a constant reminder of decline.

One summer, a small group of residents, led by retired teacher Mrs. Alvarez, decided enough was enough. They petitioned the city to let them turn the lot into something useful, a community garden. The city didn't have funds to help, but they granted permission. With borrowed tools, donated seeds, and a lot of sweat equity, the group began transforming the space.

At first, the project drew only a handful of volunteers. But as the first green shoots emerged, more people stopped by. Children helped plant sunflowers along the fence, teenagers painted bright murals, and neighbors brought folding chairs to sit and talk under the new shade of fruit trees.

Soon, the garden wasn't just about vegetables. It became a meeting place. People shared recipes, celebrated birthdays, and even started a weekly "quiet hour" where visitors could sit, meditate, or read among the flowers. The police noticed a drop in petty crime in the area. Residents reported feeling safer walking home at night. And for many, the garden became a source of calm in a stressful world, a reminder that beauty and connection could grow even in neglected places.

Mrs. Alvarez liked to say, "We didn't just plant seeds here. We planted hope."

The benefits of plants extend far beyond nutrition and oxygen, reaching deep into our emotional and psychological well-being. What was once anecdotal wisdom is now supported by decades of research: contact with green spaces can significantly improve mood, reduce stress, and even strengthen social connections. Whether in the form of a forest, a garden, or a single potted plant, nature interacts with the human mind in profound and measurable ways.

The Psychological Benefits of Being Around Plants

Exposure to green environments consistently improves mood. A 2019 meta-analysis of over 140 studies found that people who have regular contact with nature report lower levels of stress, anxiety, and depression. This effect is not limited to long hikes or wilderness trips; even small doses, such as 20 minutes in a park, can lift emotional states. Gardening, tending to indoor plants, or walking through a tree-lined street can help regulate emotions and foster a sense of calm.

Plants influence us through multiple senses. The color green, supported by color psychology research, promotes balance and relaxation. Plant aromas like lavender and rosemary have been shown in controlled studies to lower cortisol levels and reduce stress. Physical interaction, touching leaves, handling soil, or pruning, provides grounding sensory feedback that helps quiet mental chatter.

The biophilia hypothesis, proposed by biologist E.O. Wilson, suggests that humans have an innate connection to nature rooted in our evolutionary history. Cross-cultural research shows that people from all backgrounds are drawn to green spaces when given the choice. Reconnecting with plants satisfies this deep-seated psychological need, offering comfort and balance in a modern world that often pulls us away from natural environments.

Urban Greening and Social Well-being

Urban environments are associated with higher rates of stress-related illness. Studies comparing neighborhoods rich in greenery to those with minimal vegetation reveal a significant gap in life satisfaction and mental health. Even small parks or green corridors within walking distance can reduce stress and support better mental health outcomes.

Urban greening projects have been linked to reductions in crime, anxiety, and depression. These effects may be due to reduced heat stress from shaded areas, increased opportunities for social connection, and the calming visual cues provided by vegetation. Community gardens and urban farms also offer a shared sense of purpose and belonging, which strengthens neighborhood bonds.

Access to green spaces is uneven across socioeconomic lines. Disadvantaged communities often have fewer parks, fewer trees, and less safe public spaces. Inclusive urban greening policies, those that address cultural relevance, accessibility, and safety, have been shown to improve public health outcomes in cities around the world.

Plants in Schools and Workplaces

Indoor plants can improve students' concentration and memory. In classrooms, greenery paired with natural light helps students stay focused, and for children with ADHD symptoms, outdoor exposure to green environments has been shown to improve attention span, even during short activities like a 20-minute park walk.

Workplaces with plants tend to see higher productivity and lower absenteeism. Greenery improves creative thinking by up to 15% in some studies, reduces mental fatigue, and helps employees stay engaged with repetitive or high-pressure work.

Biophilic design, integrating natural elements into schools, offices, and hospitals, can reduce stress biomarkers in occupants. Living walls, rooftop gardens, and indoor greenery improve mood, create a more welcoming atmosphere, and help people feel less crowded in busy environments.

Toolkit

Here are simple ways to put this chapter into action in your own life:

1. Daily Micro-Doses of Green
• Spend 10–20 minutes outside each day, ideally in a green space like a park, garden, or tree-lined street.
• If you can't get outdoors, position yourself near a window with a view of nature or use high-quality nature imagery (landscapes, forests) on your screens.

2. Sensory Nature Reset
• Visual: Add plants to your living or working space. Aim for at least one plant within your direct line of sight from your desk or primary seating area.

• Smell: Keep a small pot of aromatic herbs like rosemary, mint, or basil, or use essential oils such as lavender in a diffuser.

• Touch: Engage in a short tactile activity like pruning a plant, touching leaves, or planting seeds to ground your nervous system.

3. Weekly Deep Nature Immersion

• Once a week, visit a larger green space (park, botanical garden, hiking trail) for at least 1–2 hours.

• Leave your phone in your pocket or on airplane mode to fully engage your senses.

4. Nature Breaks at Work

• Take short plant-based breaks every 90–120 minutes. Step outside, water an office plant, or simply gaze at greenery for 1–2 minutes to restore focus.

• If possible, position your desk near a plant cluster or window with a view.

5. Social Green Time

• Plan at least one social activity per month in a green setting, such as picnics, walking meetings, gardening with friends, or volunteering at a community garden.

• This combines the mental health benefits of nature with the protective effects of social connection.

6. Urban Greening Advocacy

• If your neighborhood lacks green spaces, join or support local greening initiatives. This could be petitioning for more trees, helping plant community gardens, or advocating for rooftop gardens.

7. Tracking and Reflection

• Keep a brief journal noting your mood, stress levels, and focus before and after your nature contact.

• After 4 weeks, review patterns; most people see measurable improvements even with small, consistent actions.

Chapter 7: Plants and Cognitive Performance

By late afternoon, Lukas often found himself staring blankly at the screen. As a freelance software developer, he thrived on solving complex coding problems. Still, lately, his brain felt like it was running on fumes by midday. The coffee on his desk was cold, and his to-do list was only half-checked.

Scrolling through an online productivity forum, Lukas stumbled on an unusual suggestion: place live plants near your workspace to improve focus. The idea sounded far-fetched, but he decided to experiment. The next day, he brought home three new companions: a tall snake plant for the corner, a peace lily for his desk, and a small rosemary bush that released a faint, invigorating aroma.

He didn't change his schedule, diet, or tools, just the greenery. Over the next few weeks, he tracked his work output. He noticed something subtle but undeniable: his afternoon slumps grew shorter, problem-solving felt more fluid, and his overall mood lifted. The rosemary's scent seemed to clear mental cobwebs, and the soft greens made his desk a more inviting place to think.

By the end of the month, Lukas had shipped two major projects ahead of schedule. The plants weren't just decoration; they were silent partners, quietly tuning his mind for better performance.

Lukas's unexpected boost in focus and output after adding plants to his workspace isn't just a personal quirk; it reflects a growing body of research on how greenery supports mental performance. Scientists are beginning to map out exactly why plants can help us think more clearly, focus longer, and even generate more creative ideas.

Understanding Attention Restoration Theory (ART)

Human attention is a finite resource. When we sustain intense mental effort, whether that's coding, studying, or problem-solving, we begin to experience what psychologists call "directed attention fatigue." This state shows up as irritability, difficulty concentrating, and slower problem-solving. In today's screen-heavy world, with constant notifications and visual demands, mental fatigue has become a near-universal experience. Attention Restoration Theory suggests that nature restores mental capacity through a phenomenon known as "soft fascination." This is when the environment gently holds your attention, like watching leaves sway or light filter through branches, without bombarding you with information. In restorative settings such as gardens, forests, or even small green courtyards, the brain can recover from fatigue. Controlled experiments show measurable improvements in memory, attention, and problem-solving after even brief exposure to nature, compared to overstimulating environments like busy streets or crowded offices. Researchers outline four key components of restorative nature experiences:

Being away: mentally and physically distancing yourself from stressors.

Extent: being in an immersive, coherent setting that feels whole and expansive.

Soft fascination: gentle engagement that refreshes rather than drains.

Compatibility: an environment that aligns with your needs, mood, and goals.

Indoor Plants and Focus

Office and Classroom Studies

Multiple studies have shown that having plants in offices or classrooms improves concentration, reduces errors, and helps people sustain attention for longer periods. Workers in plant-filled offices make fewer mistakes on reaction-time tests, and students surrounded by greenery perform better on memory and focus tasks.

Mechanisms Behind the Effect

The benefits likely come from several factors working together. Brief glances at plants during work provide "micro-restoration" moments, allowing the brain to reset without leaving the desk. Plants also improve air quality, which can subtly boost oxygen availability and may help dampen background noise perception in busy spaces.

Plant Density and Placement

The number and arrangement of plants matter. Too few, and the effect may be negligible; too many, and it can feel cluttered. Research suggests that moderate plant density, with greenery visible from most work positions, works best. Offices that arrange plants in clusters near desks or along sightlines tend to see the strongest focus improvements.

Nature doesn't just restore focus; it can also spark new ideas. Exposure to natural scenes has been linked to increased divergent thinking, a key element of creativity. Brainwave studies show shifts toward patterns associated with insight and problem-solving after time in nature. Artists, musicians, and writers have long credited gardens and green spaces as sources of inspiration.

In modern workplaces, even a few plants can play a role in innovation. Greenery provides small visual breaks that encourage the brain to make novel connections. Biophilic design, integrating natural elements into workspaces, has been shown to boost brainstorming and collaborative creativity. Innovation hubs often use plant walls or indoor trees to create a more open, idea-friendly atmosphere.

The effects of plants on cognition aren't just immediate; they can be long-lasting. Regular exposure to plant-rich environments supports mental flexibility and learning capacity over months and years. This is especially relevant for aging populations, where maintaining cognitive resilience is key to quality of life. In schools, offices, and community spaces, integrating greenery may be a simple but powerful investment in mental longevity.

Toolkit

Here are simple ways to put this chapter into action in your own life:

1. Daily Micro-Restorations

Keep at least one plant within your immediate field of vision at work or home. When you feel mental fatigue, take 30–60 seconds to look at it, noting details in its leaves, colors, and textures.

The "Green Break" Routine

Schedule a 5–10 minute walk outside every few hours, ideally somewhere with visible trees or gardens. If you can't leave the building, take a slow lap around an indoor plant area or conservatory.

2. Desk Arrangement for Focus

Place plants where you can see them without turning your head. Avoid positioning them behind you, where they lose their restorative effect. A mix of low and medium-height plants can add depth without blocking light.

3. Creative Spark Corner

Set up a dedicated space with plants for brainstorming or creative work. Include plants with varied textures and scents, such as rosemary or lavender, to subtly stimulate the senses.

Workplace Green Upgrade

If possible, introduce plants to shared spaces like meeting rooms or lounges. Aim for visible greenery in at least three directions from where people sit.

4. Mindful Plant Interaction

Once or twice a week, tend to a plant by watering, pruning, or dusting its leaves. Use the moment to slow down and let your mind wander—this often triggers fresh ideas.

Chapter 8: Plants as Climate Regulators

In the heart of a sprawling metropolis, summer was becoming unbearable. The glass-and-concrete skyline trapped heat like an oven, and residents joked grimly that you could fry an egg on the sidewalk by noon. Energy bills soared every year as air conditioners strained to keep apartments livable. Heat waves hit harder, and the air felt heavier, tinged with smog.

Then came the proposal. An architect named Lila Chen stood before the city council with an unusual plan: transform the city's flat, unused rooftops into a connected network of micro-forests. Instead of hot black tar, the roofs would host soil, native trees, pollinator-friendly shrubs, and even small ponds to collect rainwater. Skeptics called it impractical, too expensive, too heavy, too strange.

But a few property owners took a chance. The first rooftop forest bloomed atop an aging apartment block, and soon, residents noticed their units stayed cooler without blasting the AC all day. Butterflies began visiting. Birds returned. Children played on shaded patios that had once been empty, scorching hot slabs of concrete.

Within five years, the project expanded to dozens of buildings, connected by green "bridges" of vegetation. The change was measurable: temperature sensors showed that the blocks with rooftop forests were up to 7 degrees cooler in summer than surrounding neighborhoods. Air quality improved, and stormwater runoff decreased. Even crime rates in those districts fell, as residents took greater pride in their surroundings and spent more time outdoors.

Lila's rooftop forests didn't just make the city prettier; they helped it breathe, adapt, and survive.

The success of Lila's rooftop forests was more than a feel-good urban project; it was a small-scale demonstration of what scientists have long known: plants don't just make places beautiful; they actively shape the climate. From capturing carbon deep within their tissues to cooling entire neighborhoods and influencing weather patterns, plants are an underappreciated force in regulating Earth's systems. Understanding their role in climate stability is not just academic; it's essential for survival in an era of accelerating environmental change.

Carbon Sequestration — Plants as Carbon Sinks

Through photosynthesis, plants draw carbon dioxide from the atmosphere and transform it into glucose, the basic energy currency of life. This carbon becomes locked away in leaves, stems, roots, and even deep underground in soils. While all plants participate in this process, the amount and longevity of storage varies. Forests hold massive carbon reserves in their towering trunks, wetlands store it in waterlogged soils, and grasslands lock it away in deep root systems. Each ecosystem offers a unique kind of carbon vault.

Tropical rainforests act as the planet's largest and most dynamic carbon reservoirs, pulling vast amounts of CO_2 from the air each year. Boreal forests, stretching across the northern latitudes, store immense quantities of carbon in their cold, slow-decaying soils, making them long-term storage powerhouses. Preserving mature trees is critical, while saplings are vital for the future. Old-growth forests already hold centuries of accumulated carbon that cannot be quickly replaced.

When forests are cut down, their stored carbon is rapidly released into the atmosphere, intensifying climate change. Reforestation, which involves planting trees where forests once stood, and afforestation, which creates forests in previously barren areas, are among the most cost-effective climate mitigation strategies. Carbon offset programs aim to fund such efforts, though their success depends on proper implementation, long-term monitoring, and protection against future destruction.

Microclimate Control — Cooling Our Cities

Cities absorb and retain heat more than rural landscapes due to dark, heat-trapping surfaces, dense construction, and heavy traffic. This "urban heat island" effect drives up energy use, worsens air pollution, and increases heat-related health risks, especially during summer heatwaves. Tree canopies block sunlight, cooling streets and buildings beneath them. Through evapotranspiration, leaves release water vapor into the air, which lowers surrounding temperatures. Studies consistently show that tree-lined streets are measurably cooler, sometimes by several degrees, than bare concrete corridors.

Beyond street trees, cities are integrating green roofs, living walls, and expanded parks to reduce heat stress. Urban planners in many regions now set "tree canopy targets," aiming to cover a percentage of city land with leafy shade. Case studies show that after large-scale urban greening, not only do temperatures drop, but heat-related mortality declines as well.

Climate Regulation Beyond Cooling

Forests are not just passive recipients of rainfall; they help create it. Through transpiration, trees release moisture that rises into the atmosphere, condenses, and falls as rain, sustaining regional water cycles. The Amazon rainforest, for example, generates much of its own rainfall and influences weather patterns across South America. Large-scale deforestation disrupts these cycles, leading to drought in areas that once flourished.

Strategically planted trees can reduce wind speeds, shielding buildings, crops, and infrastructure. In coastal areas, mangrove forests act as living seawalls, absorbing the energy of storm surges and reducing damage to inland communities. On farmland, shelterbelts protect soil from erosion during severe winds.

Vegetation plays a central role in maintaining the balance of atmospheric gases, slowing processes like desertification, and building resilience against climate extremes. The benefits of plant-rich landscapes compound over time, cooler temperatures, richer soils, cleaner air, and a more stable climate for generations to come.

Toolkit

Here are simple ways to put this chapter into action in your own life:

1. Boost Carbon Sequestration in Your Area
• Plant long-lived, native trees whenever possible—they store more carbon over decades than fast-growing short-lived species.
• Support forest conservation groups working to protect old-growth forests, which already hold vast carbon reserves.

• Avoid products linked to deforestation, such as unsustainably sourced palm oil, beef from deforested areas, and certain tropical hardwoods.

2. Cool Your Home and Neighborhood Naturally

• Plant deciduous trees on the south and west sides of your home for summer shade and winter sunlight.

• Use climbing plants or living walls to shade exterior walls, lowering indoor temperatures naturally.

• Advocate for local policies that increase tree canopy coverage, especially in heat-prone neighborhoods.

3. Create and Maintain Green Roofs and Living Walls

• If you own or influence a building, consider installing a green roof—these reduce heat absorption, improve insulation, and provide habitat.

• Living walls can be built indoors or outdoors, cooling spaces while also filtering air.

4. Support Local Rainfall Cycles

• Plant trees and shrubs in clusters to maximize transpiration and moisture release into the air.

• Reduce paved surfaces where possible to allow rainwater to soak into the soil, supporting healthier vegetation.

• Join community tree-planting initiatives to expand regional vegetation cover.

5. Protect Against Wind and Storm Damage with Plants

• In rural or coastal areas, plant windbreaks or hedgerows to protect crops and buildings.

• Support mangrove restoration projects in coastal zones—they provide some of the strongest natural defense against storms.

6. Choose Climate-Positive Landscaping

• Favor native species—they thrive with less water and maintenance, and provide habitat for local wildlife.

• Reduce lawn space in favor of mixed plantings of shrubs, flowers, and trees, which store more carbon and cool more effectively.

7. Take the Long View

• Remember that the climate benefits of planting are cumulative. The earlier you plant, the more years of cooling, carbon storage, and climate stabilization your community gains.

• Encourage intergenerational stewardship—teach younger people how to plant and care for vegetation so benefits last for decades.

Chapter 9: Water, Soil, and Air Quality

Luis had lived on his family's farm for as long as he could remember. The river that wound through the valley had been the lifeblood of the land, supplying irrigation, supporting fish, and attracting birds that kept pests in check. But over the years, he had watched it shrink to a muddy trickle. Without the cool shade of trees, the water warmed, algae bloomed, and the fish disappeared. Seasonal rains washed bare soil into the stream, turning it cloudy and choking what life remained.

Most of his neighbors assumed the decline was permanent, a result of climate change and bad luck. But Luis remembered his grandfather's stories of a time when willows lined the banks, when wildflowers covered the meadows, and when the water stayed clear even in dry months. He decided to try something different, not with machines or chemicals, but with plants.

He began by planting native trees along the banks: alder, willow, and cottonwood. Their roots dug deep into the soil, anchoring it against erosion. He added shrubs and grasses in between, creating a living buffer that filtered runoff before it reached the water. At first, the changes were slow. But after two rainy seasons, something shifted. The water stayed clearer, the banks stopped collapsing, and frogs began to sing at night. Within five years, the stream had deepened and widened naturally. The shade from the trees cooled the water, bringing back trout and other species that had been absent for decades. Birds nested in the branches, and beavers returned, building small dams that created new wetland pockets. Luis's neighbors noticed that his irrigation water lasted longer through the summer, and they began planting their own riparian buffers. What started as one farmer's quiet experiment spread throughout the valley. The river became a shared treasure again—a reminder that sometimes the best way to heal water is to plant the right life around it. Luis's success in bringing life back to the river wasn't just about beauty; it was about function. His trees, shrubs, and grasses became more than scenery; they acted as engineers, stabilizing soil, filtering water, and refreshing the air. What happened on his farm illustrates a much larger truth: plants are silent guardians of Earth's most essential resources, soil, water, and air.

Plants as Guardians of Soil

Plant roots bind soil particles together, reducing the amount of sediment washed away by rain or blown off by wind. By holding the land in place, vegetation helps prevent devastating erosion events that strip away fertile topsoil. Forested slopes and grass-covered hillsides act like living armor, slowing water runoff and absorbing its force. In areas prone to landslides, deep-rooted trees provide crucial stability, while agricultural windbreaks and contour planting redirect wind and water in ways that protect crops and land alike.

Leaf litter, fallen branches, and decomposing plant matter enrich topsoil with organic nutrients. As roots grow and die, they create tiny channels that allow rainwater to penetrate deeper, improving water infiltration and storage. Cover crops, plants grown primarily to protect and enhance soil, play a vital role in sustainable farming by preventing erosion, fixing nitrogen, and fostering beneficial soil microbes.

In dry, degraded landscapes, planting trees and native shrubs can halt or even reverse desertification. Their roots anchor shifting sands, while their shade reduces soil temperature and evaporation. Large-scale projects like Africa's Great Green Wall demonstrate the power of vegetation to restore life to once-barren regions. Choosing native species is essential because they are adapted to the local climate and can thrive with minimal water, making them more sustainable in the long term.

Plants in Water Quality Protection

Wetlands act as Earth's natural water treatment systems. The plants within them absorb and trap excess nutrients, heavy metals, and sediments, purifying water before it flows downstream. Even in urban environments, constructed wetlands are being used to manage stormwater runoff, filtering pollutants from roads and buildings before they reach rivers and lakes.

Through transpiration, plants release water vapor into the atmosphere, contributing to cloud formation and rainfall patterns. Forests function as "natural water towers," storing moisture in vegetation and soil, then gradually releasing it to sustain streams and rivers during dry periods. When vegetation is lost, this natural regulation breaks down, intensifying droughts and disrupting rainfall cycles.

Strips of vegetation along riverbanks known as riparian buffers act as living filters. They trap sediments, absorb excess nutrients, and reduce the flow of pollutants into waterways. These buffers not only protect water quality but also provide shade that keeps streams cool for fish, stabilize banks, and create habitat for wildlife. Restoration projects around the world have shown that replanting these critical zones can rapidly improve aquatic biodiversity and fish populations.

Plants and Air Purification

In the late 1980s, NASA conducted research to identify plants capable of removing volatile organic compounds (VOCs) from sealed environments like space stations. Plants such as the peace lily, spider plant, and snake plant were found to filter certain indoor air pollutants effectively. While groundbreaking at the time, these results were based on controlled, small-scale conditions.

In everyday environments, the number of plants required to significantly purify indoor air is far greater than most homes or offices contain. Good ventilation remains the most effective way to maintain healthy air indoors. However, plants still serve as a valuable supplement, improving humidity, providing minor filtration, and contributing to a healthier environment.

Even when their direct impact on air chemistry is limited, plants can make the air feel fresher simply through sensory perception. The presence of greenery can lower stress, promote relaxation, and encourage a sense of well-being. Combined with other green building strategies, plants create spaces that not only support physical health but also nourish mental and emotional wellness.

Toolkit

Here are simple ways to put this chapter into action in your own life:

1. Choose Plants for Your Local Environment
   Select native or well-adapted species for your climate and soil type. These plants will thrive with less water, fertilizer, and maintenance, making them more sustainable over time.
2. Prevent Soil Erosion at Home
   Plant groundcovers, grasses, or shrubs on bare soil to anchor it in place. On slopes, use terracing or contour planting to slow water flow and reduce runoff.
3. Enrich Soil Naturally
   Add compost, mulch, and leaf litter to improve soil structure and nutrient content. Rotate crops in gardens and use cover crops such as clover or rye to prevent nutrient depletion.
4. Create a Mini "Water Filter" Zone
   If you have a property with drainage areas or ditches, plant reeds, cattails, or moisture-loving shrubs along the path of water flow. These plants trap sediment and absorb pollutants before water leaves your property.
5. Support Riparian Buffers
   If you live near a stream, pond, or lake, keep a strip of undisturbed vegetation along the bank. Avoid mowing right to the water's edge and plant native grasses or shrubs to prevent erosion.
6. Bring Air-Purifying Plants Indoors
   Add hardy, low-maintenance species such as peace lily, snake plant, or pothos to workspaces and living areas. While they won't replace ventilation systems, they can help with humidity balance and create a fresher-feeling environment.
7. Reduce Air Pollutants Outside
   Plant trees or shrubs along property lines to act as windbreaks and reduce dust. In urban settings, support community tree-planting programs to improve neighborhood air quality.
8. Combine Greenery with Other Eco Practices

Pair plant-based solutions with other strategies: rain barrels to reduce runoff, composting to reduce waste, and natural ventilation to complement indoor plants.

Chapter 10: Biodiversity Support

In the sunbaked hills of Andalusia, the village of San Javier had been farming olives, almonds, and vegetables for generations. The air once hummed with the sound of bees in spring, but over the years, the buzzing faded into silence. Farmers still worked their fields, but their almond harvests came in light, and vegetables struggled to set fruit.

At first, the villagers blamed unpredictable rains and hotter summers. It wasn't until María, a young agronomy student who had grown up there, came home for a summer break that someone connected the dots. She noticed that the hedgerows between fields had been cleared to make room for more planting, and wildflower patches were gone. Years of pesticide use had also left their mark.

María spoke with the elders and explained that their missing pollinators, bees, butterflies, and other small creatures, were the invisible workers behind their past abundance. Without them, the plants couldn't reproduce properly. Her words stirred a quiet urgency.

A small group of residents decided to try an experiment. They replanted native wildflowers along the dirt paths and field edges. They allowed patches of weeds, many of which were actually medicinal herbs, to grow. Olive groves were interspersed with flowering shrubs, and chemical sprays were replaced with organic pest control. Children helped build bee hotels from hollow canes and old wood, while the mayor secured a small grant to fund more planting.

By the second spring, splashes of color painted the once-barren edges of the fields. Bees returned first, then butterflies, then swallows swooping to catch the insects in flight. That year, the almond trees were heavy with fruit. The vegetables were healthier, and even the olive harvest was stronger.

Word of the transformation spread. An ecotourism company began bringing visitors to see the wildflower corridors and taste local honey produced by new village hives. The economy lifted not just from better harvests, but from the visitors eager to spend a day in the "bee village." For the people of San Javier, the change wasn't just economic. The hum of bees had returned, and with it, a sense that they were again part of a living, breathing landscape that they had the power to protect.

The revival of San Javier's pollinator population was more than a local success story; it was a vivid example of how plants act as the backbone of biodiversity. When the villagers brought back native flowers, they didn't just help bees; they restored an entire web of life, from the insects that pollinated crops to the birds that fed on them. The lesson was clear: protecting and diversifying plant life is not just an act of conservation, it's an investment in the health, stability, and prosperity of entire ecosystems.

Plants as the Foundation of Ecosystems

Plants serve as the structural framework of nearly every terrestrial ecosystem. Trees, shrubs, grasses, and mosses all create shelter for countless species of insects, birds, and mammals. From towering canopy layers that house birds of prey to dense understory plants that provide cover for small mammals, and lowlying ground covers that hide insects and amphibians, plant life creates layered habitats that support diverse life forms. Seasonal changes in plant communities, spring blossoms, summer foliage, autumn fruits, and winter seed heads offer year-round resources for wildlife.

All life in an ecosystem depends on the energy captured by plants through photosynthesis. This energy flows from plants to herbivores, and then up the food chain to predators. Plants also maintain mutualistic relationships with fungi and microbes, exchanging nutrients in ways that sustain soil health and biodiversity. When plant diversity declines, the effects ripple outward, disrupting pollinators, herbivores, predators, and decomposers alike.

Biodiverse plant systems are more resilient to disease, pests, and environmental stress. When a variety of plant species is present, it supports a wider range of animals, microorganisms, and fungi, creating a robust system where no single disturbance can collapse the entire community. Tropical rainforests and prairie grasslands are prime examples, where high plant diversity ensures long-term ecological stability.

Supporting Pollinators and Food Crops

Many plants rely on specific pollinators, bees, butterflies, birds, and even bats, for reproduction. Plants evolve colors, scents, and flower shapes to attract these partners, forming intricate ecological relationships. Some crops, such as almonds and cacao, are almost entirely dependent on wild pollinators for their yields. Without these partnerships, food production and biodiversity suffer.

Habitat loss, pesticide use, and climate change are all driving declines in pollinator populations. The disappearance of bees, in particular, poses a serious risk to global agriculture. Monoculture farming also contributes to the problem, reducing the diversity of nectar and pollen sources needed to sustain healthy pollinator communities year-round.

Pollinator recovery efforts focus on restoring native plant habitats, creating flowering corridors that connect fragmented landscapes, and supporting urban beekeeping. Pollinator-friendly farming practices, such as reducing pesticide use and planting wildflower borders, can significantly boost biodiversity. Public policies and incentive programs have also proven effective in encouraging landowners to prioritize pollinator habitat restoration.

The Economic Value of Ecosystem Services

Ecosystem services are the benefits that nature provides to humans. Provisioning services include food, timber, and medicinal plants; regulating services involve climate control and water purification; and cultural services encompass recreation, aesthetic enjoyment, and spiritual enrichment.

Pollination alone is valued at billions of dollars globally each year. When biodiversity declines, the economic losses can be enormous, affecting agriculture, fisheries, tourism, and even property values. However, assigning monetary value to nature's benefits is controversial; some argue that it reduces complex ecosystems to mere commodities.

Restoring plant diversity has shown measurable economic returns, from increased crop yields to growth in ecotourism. Governments and corporations are beginning to make biodiversity commitments, investing in green infrastructure, habitat restoration, and sustainable land management. Protecting plant-rich ecosystems isn't just an environmental imperative; it's a long-term strategy for economic and ecological resilience.

Toolkit
Here are simple ways to put this chapter into action in your own life:

1. Plant Native Species
Choose plants that are naturally adapted to your region. They require less maintenance, provide food for local wildlife, and help restore natural ecosystem balance.
2. Create Habitat Layers
In gardens or landscapes, mimic natural ecosystems with canopy trees, shrubs, ground covers, and flowering plants. Each layer supports different species.
3. Support Pollinators
Plant a variety of flowers that bloom at different times of year to provide continuous food for bees, butterflies, and other pollinators. Avoid pesticide use, especially during bloom periods.
4. Build Wildlife Corridors
If you have space, create green pathways between garden areas or across properties. This allows animals and insects to move safely between feeding and nesting areas.
5. Encourage Urban Biodiversity
In cities, support rooftop gardens, green walls, and pocket parks. Even small green spaces can help connect fragmented habitats and support pollinators.
6. Reduce Lawn Space
Replace parts of your lawn with wildflowers, shrubs, or vegetable gardens. Lawns often provide little to no food or shelter for wildlife.
7. Get Involved in Local Restoration Projects
Join or support groups that plant trees, restore wetlands, or remove invasive species in your area. Collective action has far-reaching ecological benefits.
8. Educate and Advocate

Share information about the importance of biodiversity with friends, neighbors, and community groups. Encourage local leaders to implement pollinator-and habitat-friendly policies.

9. Support Sustainable Farming and Forestry

Choose products from farms and forests that prioritize biodiversity. Certifications like Rainforest Alliance or organic labeling can help guide your choices.

10. Monitor and Celebrate Biodiversity Gains

Track the wildlife and plant species you see in your area over time. Celebrate the return of pollinators, birds, and other species—it keeps motivation high and shows the impact of your efforts.

Chapter 11: Plants as Economic Engines

In the mountains of Nara Prefecture, Japan, lies the small town of Yoshino, a place where nature and economy are woven together as seamlessly as silk. For over 1,300 years, the people of Yoshino have cultivated cherry blossom trees along the slopes that rise above the town. What began as a cultural and spiritual tradition slowly evolved into one of the most remarkable examples of nature-based economic planning in the world.

Each spring, as winter retreats, Yoshino transforms. The mountainsides ignite in cascading waves of soft pink and white. The sight draws nearly 200,000 visitors from across Japan and around the globe in just a few weeks. Hotels are booked months in advance, and restaurants serve seasonal menus infused with cherry blossom flavors: sakura tea, sakura mochi, even cherry blossom salt for sprinkling over rice.

The economic impact is staggering for such a small town. Local artisans craft limited-edition woodblock prints, hand-painted ceramics, and textiles patterned with blossoms. Farmers sell blossom-themed sweets, pickles, and preserves. Tour guides, photographers, and performers find work during the bloom season, all tied to the fleeting beauty of these trees.

But behind the seasonal spectacle lies year-round dedication. The residents of Yoshino care for the trees meticulously, pruning, planting, and protecting them from disease. They have mastered the art of succession planting so that when older trees reach the end of their life cycle, new saplings are ready to take their place. This ensures that the "blossom economy" never runs out of its most valuable resource.

The blossoms are more than just flowers; they are an economic engine, a cultural identity, and a symbol of harmony between human livelihood and the natural world. In Yoshino, the lesson is clear: when nature thrives, so do the people.

The story of Yoshino shows us that plants are not just a backdrop to human activity; they are drivers of prosperity, culture, and long-term stability. Whether they grow in a small mountain town or across vast agricultural plains, plants form the basis of countless industries that keep societies thriving. From feeding billions to inspiring billion-dollar tourism seasons, their economic role is as diverse as it is essential.

The Pillars of the Global Plant Economy

At the heart of the plant economy is agriculture, the most fundamental and widespread form of human reliance on plants. Crops like rice, wheat, and maize feed billions of people every day, forming the dietary staples of much of the world. These crops not only sustain lives but also underpin global trade, with agricultural products accounting for a significant share of many countries' GDP. Shifts toward plant-based diets and alternative protein sources are now reshaping this sector, as consumer demand grows for foods that are both sustainable and health-conscious. This transition is sparking innovation across farming, food production, and distribution systems worldwide.

Beyond the dinner table, plants fuel a vast horticultural industry. Landscaping, ornamental flowers, and indoor greenery are all part of a thriving global market that influences everything from urban planning to holiday traditions. Seasonal spikes in demand, such as cut flowers for weddings, poinsettias at Christmas, and cherry blossoms for spring festivals, demonstrate the powerful intersection of culture and commerce. These plants may not feed the body, but they feed economies and cultural identity alike.

Forests contribute not only timber for construction but also a wealth of non-timber products, including resins, nuts, and medicinal plants. When managed sustainably, forestry can be a renewable economic pillar, balancing extraction with regeneration. Certification programs like FSC (Forest Stewardship Council) and PEFC (Programme for the Endorsement of Forest Certification) guide this balance, ensuring that wood and forest products are sourced responsibly to preserve ecosystems for future generations.

Ecotourism and Nature-Based Industries

Nature-based tourism is a growing sector that depends heavily on plant life to attract visitors. Rainforest treks, desert wildflower blooms, and botanical gardens draw millions annually, generating income for local communities and funding conservation projects. Seasonal plant spectacles like cherry blossom season in Japan or fall foliage in New England can become central pillars of regional economies.

Successful ecotourism hinges on preserving the very plant life that draws travelers. In Costa Rica, rainforest tourism thrives because of strict conservation policies. In Holland, vast fields of tulips not only attract millions of visitors but also drive an entire export industry. Community-based tourism projects that protect plant ecosystems often generate more sustainable income than industries that degrade them, proving that protecting biodiversity can be profitable.

However, when plant-based attractions become too popular, overtourism can threaten their survival. High visitor numbers can damage fragile habitats, leading to long-term losses in both biodiversity and income. The solution lies in balancing access with protection through visitor limits, education programs, and infrastructure designed to minimize environmental impact.

Emerging and Innovative Plant-Based Industries

Plants are also stepping into the spotlight as the foundation for next-generation materials. Plant-based plastics and biodegradable packaging offer eco-friendly alternatives to petroleum-based products. Hemp and bamboo are being used in sustainable construction, while algae and crop waste are fueling advances in biofuels. These industries are set to grow as global demand for sustainable solutions accelerates.

Medicinal plants remain a cornerstone of both traditional healing and modern pharmaceuticals. Ongoing research into plant compounds is leading to new drugs, functional foods, and health supplements. The global nutraceutical market is expanding rapidly, driven by consumer interest in natural, plant-derived wellness products.

The future of the plant economy includes vertical farming and urban agriculture, which bring food production into cities and reduce transportation costs. Climate-smart agriculture, which integrates resilience and sustainability into farming practices, is attracting significant investment. Large-scale reforestation offers another high-potential economic opportunity, promising both environmental and financial returns as carbon markets expand.

When viewed together, these industries demonstrate a truth as clear as the blossoms of Yoshino: plants are not just part of the economy, they are the economy's roots, branches, and seeds for the future.

Toolkit

Here are simple ways to put this chapter into action in your own life:

1. Identify Local Plant-Based Industries

Walk through your community and note the plant-related industries already present—farms, nurseries, landscaping services, herbal shops, woodworking businesses, or ecotourism ventures. Mapping them helps reveal economic connections you may not have noticed.

2. Support Sustainable Agriculture and Forestry

When buying food, choose locally grown, seasonal produce to keep money circulating within your region. For wood products, look for FSC or PEFC certifications that indicate responsible forestry.

3. Explore Plant-Based Business Opportunities

If you're an entrepreneur, consider emerging markets: plant-based packaging, hemp construction materials, herbal supplements, or vertical farming. These sectors are growing rapidly and often have lower environmental footprints.

4. Connect Conservation with Commerce

If you work in tourism or community development, design programs that link plant conservation with visitor experiences—such as guided nature walks, botanical festivals, or native plant gardens that educate and entertain.

5. Invest in Community Green Infrastructure

Encourage or participate in projects like community gardens, green roofs, and tree-planting initiatives. These improve the quality of life while creating economic benefits through energy savings and increased property values.

6. Partner with Local Growers and Artisans

Restaurants, cafes, and retailers can strengthen the plant economy by sourcing ingredients, flowers, and decorative plants from local growers. This fosters resilience and brand authenticity.

7. Advocate for Plant-Centered Policy

Support legislation that protects plant habitats, funds sustainable agriculture research, or incentivizes green technology. Public policy often shapes the long-term health of plant-based industries.

By taking these actions, you're not only supporting your local economy, you're also reinforcing the living systems that make prosperity possible. Every seed planted, tree preserved, and plant-based product chosen strengthens the roots of economic and ecological stability.

Chapter 12: Mutual Benefit

Far above the Arctic Circle, where winter nights last for months and the air is sharper than glass, there's a steel door buried in the side of a frozen mountain. It leads into a hallway lit with pale, cold light. At the end of that hallway is the Svalbard Global Seed Vault, one of humanity's most essential and least visited treasures.

Inside, millions of seeds from almost every nation on Earth rest in sealed packets, locked away in sub-zero storage. Each one represents a piece of our agricultural heritage: rice from Asian paddies, wheat from Middle Eastern fields, maize from the Americas, and millet from African plains. They are kept here not because the world needs them now, but because one day, we might.

The vault has already proven its value. When the war in Syria destroyed the seed bank in Aleppo, scientists turned to Svalbard. They requested duplicates of seeds they had once deposited, and from this frozen mountain, life returned. Seeds shipped back, planted again, and grown into crops that could restore a nation's food supply.

This is more than storage; it's a living insurance policy, one that depends on global cooperation. Farmers, scientists, and governments agree to send their most precious seeds here, trusting that if disaster strikes (whether it's a changing climate, a devastating pest, or political unrest), this vault will be the bridge between loss and recovery.

Human cultivation has always been about more than feeding ourselves. It's about giving plants a chance to thrive far beyond their native habitats, preserving their diversity, and ensuring their survival for future generations. And nowhere is that mission more literal, or more urgent, than in this silent, frozen chamber at the edge of the world.

The story of the Svalbard Global Seed Vault is a reminder that our relationship with plants is not a one-way exchange. We take from them, yes, food, medicine, shelter, beauty. Still, we also have the ability (and the responsibility) to safeguard them in return. This chapter explores the many ways humans have become caretakers of plant life, helping species travel beyond their native ranges, preserving their genetic diversity, and creating systems that allow both plants and people to thrive together.

How Human Cultivation Helps Plants Thrive Beyond Their Native Habitats

Human history is also the history of plants on the move. Wheat, first domesticated in the Fertile Crescent, now grows in fields from Canada to Australia. Potatoes, once confined to the cool slopes of the Andes, became a staple across Europe and Asia. These journeys were made possible through trade routes, migrations, and global exploration pathways that carried seeds, cuttings, and tubers into new climates and soils. Over generations, cultivated plants adapt, shifting their growing seasons, altering their resilience to pests, and even developing new traits that make them better suited to their adopted homes. In this way, human cultivation has extended plant ranges far beyond what natural dispersal could have achieved.

Beyond farms and fields, humanity has created specialized sanctuaries for plant life. Botanical gardens serve as living museums, showcasing plants from around the globe while preserving rare, endangered, or even extinct species in the wild. These gardens also function as research hubs, where scientists experiment with breeding and hybridization to strengthen plants against disease, drought, and changing temperatures. There are notable success stories of species that were on the brink of extinction saved through careful cultivation and reintroduction to their native habitats. In these cases, gardens are not just preserving beauty; they are acting as life-support systems for the plant kingdom.

While moving plants beyond their native habitats can be beneficial, it comes with risks. Introduced species can sometimes become invasive, outcompeting native plants and disrupting entire ecosystems. History is filled with examples of well-intentioned introductions that ended in ecological imbalance. That's why stricter protocols, including international regulations on imports and exports, guide plant movement today. These frameworks aim to balance the benefits of cultivation and trade with the responsibility to protect native ecosystems, ensuring that our global garden remains a place of cooperation rather than competition.

Seed Banks and Conservation Programs

Every seed holds a library of genetic information, and that diversity is the foundation of a plant's ability to survive change. Without it, species become vulnerable to pests, diseases, and shifts in climate. Seed banks serve as vaults for this genetic wealth. Facilities like the Svalbard Global Seed Vault in Norway store duplicates of seeds from around the world, acting as a backup for humanity's most essential crops. National seed banks, maintained by individual countries, safeguard varieties important to local agriculture and ecosystems. Together, these collections act as a form of insurance against future crises, whether they are natural disasters, political conflicts, or environmental shifts.

Building a seed bank is a precise and patient science. Seeds are collected from healthy, genetically representative plants, then carefully cleaned to remove debris and pests. They are tested for viability to ensure they can germinate in the future. Storage often takes place in controlled environments with low humidity and low temperature to slow the seed's natural aging process. Periodically, seeds are removed, germinated, and grown into new plants so that fresh seeds can be harvested and returned to storage. This cycle ensures the genetic material remains alive and ready for use decades, or even centuries, later.

Preserving seeds is not a task that can be done in isolation. International treaties, such as the International Treaty on Plant Genetic Resources for Food and Agriculture, facilitate the exchange of seeds between countries. Non-governmental organizations, universities, and farming communities all contribute to this global effort. At the local level, community seed-saving initiatives empower farmers to preserve traditional crop varieties that may not be represented in formal seed banks. Together, these efforts form a global safety net for plant diversity, linking conservation with food security for generations to come.

The Role of Sustainable Agriculture in Protecting Both Plants and People

Not all agriculture depletes the environment; when done thoughtfully, it can enhance it. Practices like crop rotation, intercropping, and agroforestry create agricultural landscapes that mimic natural ecosystems. Organic and regenerative farming approaches minimize pesticide and chemical fertilizer use, helping protect pollinators, maintain soil health, and encourage beneficial microorganisms. By treating the farm as part of a larger ecosystem, these methods turn fields into shared spaces where plants, wildlife, and people can thrive together.

The wild relatives of domesticated crops are an overlooked treasure. They often carry genetic traits such as drought tolerance, disease resistance, or nutrient efficiency that could prove invaluable for future farming. Protecting these wild species, whether in nature reserves or integrated into farming landscapes, preserves a vital source of resilience. Plant breeders can cross-cultivate varieties with their wild counterparts to create hardier crops capable of thriving in an unpredictable climate.

At the core of sustainable agriculture is the goal of food security, not just for today, but for decades to come. When farms protect plant diversity, they create buffers against environmental and economic shocks. This stability benefits rural communities, providing a steady income and reducing reliance on external inputs. Aligning agricultural policy with plant conservation goals ensures that farmers are supported in making choices that benefit both the land and the marketplace. In this way, agriculture becomes more than food production; it becomes a partnership between humans and plants, each safeguarding the other's future.

Toolkit

Here are simple ways to put this chapter into action in your own life:

1. Start a Local Seed-Saving Group
   Gather neighbors, community gardeners, and local farmers to exchange seeds from your own crops or native plants. This grassroots approach preserves varieties adapted to your local climate while strengthening community ties.
2. Visit and Support Botanical Gardens
   Botanical gardens often fund conservation and research through ticket sales and donations. Visiting these spaces helps sustain their work in preserving rare and endangered species, while giving you inspiration for your own plantings.
3. Choose Native and Region-Adapted Plants for Your Garden

When selecting plants, prioritize species native to your area or proven to thrive in your climate without heavy inputs. This reduces water use, supports local pollinators, and strengthens your region's biodiversity.

4. Volunteer for a Seed Bank or Conservation Program
   Look for opportunities at local universities, agricultural research stations, or community-led seed libraries. Volunteers can assist with seed collection, cleaning, labeling, or educational outreach.

5. Practice Crop Diversity in Your Own Growing Space
   Whether you have a small garden or a large farm, mix crops to reduce pest pressure, improve soil health, and create microhabitats for beneficial insects.

6. Support Policies that Protect Plant Diversity
   Engage with local policymakers or agricultural boards to advocate for programs that promote native species restoration, protect wild plant relatives, and encourage sustainable farming.

7. Learn About Wild Crop Relatives in Your Region
   Identify wild species connected to domesticated crops and understand their habitats. If possible, support initiatives that protect these plants through land conservation or responsible foraging.

8. Reduce Reliance on Single-Source Foods
   In your diet, incorporate a greater variety of grains, vegetables, and fruits to encourage market demand for diverse crops. This supports farmers who grow beyond the typical staples and helps keep rare varieties in production.

9. Avoid Introducing Invasive Species
   Before planting or importing ornamental or exotic plants, check your region's prohibited species list to ensure they won't escape cultivation and disrupt local ecosystems.

10. Share Knowledge About Sustainable Agriculture
    Host workshops, post on social media, or give talks at schools or gardening clubs to spread awareness about how human cultivation and conservation can work together for mutual benefit.

Chapter 13: Beyond Science — More Than Meets the Eye

Joe had come to the Andes chasing data, not folklore. My grant proposal was clear: study soil conditions and microclimates in high-altitude maize farming. But in the first week, something happened that no spreadsheet could prepare him for.

It was dawn when Don Mateo, a wiry man with deep lines etched by sun and wind, motioned for Joe to follow him into the fields. The stalks from last season stood dry and brittle, their faded husks clinging like ghosts to the stems. Joe expected him to check the soil or test the temperature, but instead, he closed his eyes.

They stood there in silence, the chill of morning biting at their fingertips. Then the wind shifted softly, slowly, and passed through the rows with a faint rustle. It was nothing remarkable to me, just the ordinary sigh of plants in a breeze. But Mateo nodded as if he'd heard something precise, something important.

"It is time," he said simply.

Later, Joe learned this was how his people had planted for centuries, waiting for the sound of the wind in the maize to change, a signal they swore meant the earth and sky were ready to receive new seeds. No thermometer, no calendar, no lab test, just listening.

Joe thought it was romantic nonsense. But when the planting was done and the months passed, their fields thrived while neighboring ones, planted earlier or later, struggled in patchy rains.

Even now, Joe can still remember that morning's rustle. Part of hi insists it was a coincidence. Another part wonders if he heard it too, but didn't yet know how to listen.

The moment in the field left him with more questions than answers. It was not proof of anything, yet it lingered in his mind in a way that numbers and charts never had. That's the nature of certain encounters: they resist dissection, but still leave a mark. Across history, countless people have described similar moments of fleeting instances when the plant world seems to step forward and meet us halfway.

These moments are where the boundaries of science blur into the terrain of personal experience, where knowledge is carried not only in books and studies but in memory, story, and intuition.

A Story That Opens the Door

In the Amazon, some shamans say their plant knowledge did not come from experimentation alone, but through dreams. In these visions, the spirit of a plant would teach its uses, sometimes offering detailed instructions on preparation and healing. Such accounts might be dismissed as myth until modern science later confirms their accuracy, as happened with quinine from cinchona bark, used for centuries to treat fevers before its antimalarial properties were documented.

Elsewhere, in ancient China, farmers observed the blossoming of certain flowers as a sign that seasonal rains were coming, timing their planting accordingly. In Polynesia, navigators used the flowering cycles of coastal shrubs to predict changes in the sea part of a sophisticated knowledge system that guided long voyages without modern instruments.

These practices were not "superstitions" to their practitioners; they were practical, tested, and deeply embedded in daily life.

Think back, have you ever had a moment where a plant seemed to meet you in some unexplainable way? Perhaps you found comfort in a tree you liked to sit under as a child. Maybe you've noticed a houseplant seeming to perk up when you're in the room, or a wildflower appearing at the very moment you needed beauty most.

You may have brushed it aside, yet such experiences have been recorded in every culture and every generation. They may not be measurable, but their meaning is undeniable to the person who feels them.

In the chapters ahead, we will explore these moments not to prove or disprove them, but to understand them as part of a long human–plant relationship that is as much emotional and intuitive as it is biological.

Why Explore the Unseen Side of Plants?

Science has given us extraordinary tools for understanding plants, microscopes to see their cells, sensors to measure their growth, and chemical tests to identify their compounds. Yet it struggles to measure subjective experience: the feeling of connection, the sense of being "noticed," the intuition that guides a farmer or healer.

There are countless plant-related phenomena that sit in this uncharted space described in stories, repeated across continents, yet not easily replicated in a lab. Rather than dismiss them, we can approach them with curiosity, acknowledging that our instruments may not yet be tuned to detect all that is happening.

Across the globe, traditional knowledge systems have long described plants as more than resources. In Ayurveda, herbs are categorized not only by their chemical effects but by their energetic qualities: warming, cooling, and grounding. In Native North American traditions, plants like sage and sweetgrass are used for cleansing the spirit as well as the body. These uses often predate scientific verification by centuries. For example, lavender's calming reputation was once considered folklore; today, it is recognized for containing compounds that ease anxiety. Again and again, what begins as traditional wisdom often finds confirmation in the lab, sometimes long after its benefits have been known in practice.

History shows that science and spirituality are not enemies but partners in discovery. Concepts once dismissed as fantasy, like plants releasing airborne chemicals to warn each other of pests, are now well documented.

By holding both perspectives, we can create a fuller picture: science offers precision and proof, while traditional knowledge offers context, depth, and centuries of lived experience. Together, they can reveal more than either could alone.

Acknowledging the Nature of This Journey

Some aspects of the plant-human connection will remain personal, unrepeatable, and unmeasurable. But this does not mean they lack value. Agricultural practices like crop rotation were effective for centuries before science could explain why they worked.

Cultural memory preserved in stories, rituals, and seasonal observances can carry truths that data alone cannot capture. Sometimes, it is worth listening to knowledge even when we cannot yet prove it.

Belief shapes perception. When you believe that plants can communicate in some way, you begin to notice subtle shifts in the scent of a flower, the posture of leaves in your presence. Skeptics may see a coincidence; others may see a relationship.

Regardless of interpretation, the benefits of engaging with plants in a conscious, even spiritual way are clear. It can reduce stress, promote well-being, and strengthen a sense of belonging to the living world.

Setting Expectations for the Reader

As we move forward, you will encounter both scientifically validated research and accounts rooted in tradition, experience, or speculation. Some will be familiar; others may challenge what you believe is possible. You are not being asked to accept every claim as fact. Instead, take what resonates with you and let the rest be seed ideas that may grow in meaning over time.

The purpose here is not persuasion, but expansion: widening the scope of what plant–human relationships might mean, and honoring the possibility that the green world may be more aware, more connected, and more engaged with us than we yet understand.

Toolkit

Here are simple ways to put this chapter into action in your own life:

1. Observation Journal

Keep a small notebook (or digital notes app) specifically for plant encounters. Over the next two weeks, record any moments, no matter how subtle, when you feel a plant draws your attention, affects your mood, or changes in your presence. Include details like:

• Date, time, and location
• The plant species (if known)
• Weather, light, and surrounding sounds or scents
• Your emotional or physical state before and after the moment

2. Sensory Immersion Exercise

Choose one plant to spend at least five minutes with each day for a week. Sit or stand quietly nearby and focus on one sense at a time:

• Sight — Study the shapes, colors, and patterns in detail.
• Sound — Listen for movement in leaves, stems, or nearby wildlife.
• Smell — Inhale gently to notice scent changes at different times of day.

• Touch — (If safe) feel the texture of leaves, bark, or petals.

Write down any impressions or feelings that arise.

3. Cross-Cultural Knowledge Search

Pick a plant you encounter often (e.g., oak, lavender, aloe). Research how at least three different cultures have traditionally used or understood this plant. Compare these uses to modern scientific findings.

Ask yourself: Do traditional stories reveal something science has only recently discovered or has yet to?

4. Intuitive Connection Practice

Before watering or tending to a plant, pause for a slow breath. Imagine "greeting" the plant in your mind. Notice if anything changes in your perception, perhaps a shift in the plant's posture, your own body's relaxation, or a new thought about its care.

5. Dialogue with a Plant

This may feel unusual at first, but many cultures use some form of this practice. Choose a plant and quietly "ask" it a question (about its needs, about your own life, or about the space you share). Be open to receiving an answer in the form of a sudden thought, memory, or image. Record the experience without judgment.

6. Skeptic's Balance Method

If you lean toward skepticism, choose one plant-related tradition or claim that interests you but seems unlikely. Research both the cultural history and the current scientific perspective. Look for overlaps, any shared observations or results. This can help bridge intuition and analysis.

7. Reflect and Share

At the end of the month, reread your journal. Highlight entries that feel significant or mysterious. If you feel comfortable, share one of these moments with a friend, family member, or plant-focused community group. Often, hearing others' experiences helps you recognize patterns in your own.

Chapter 14: The Concept of Plant Consciousness

It's 2140, and I am standing on a quiet balcony thirty stories above the heart of New Nairobi. The skyline isn't a jagged forest of concrete and glass anymore; it's alive. Vines drape between buildings like suspension bridges, their tendrils thick and coiled with purpose. Towering baobabs pierce the cityscape, their roots winding down through hollow elevator shafts into nutrient-rich soil beds deep below.

A soft chime from my wristband announces that the "Council" is ready. Inside, in a chamber kept at the temperature of a warm tropical dawn, six massive plants sit in a semi-circle, their leaves trembling in response to subtle changes in the air. At the center is Aiya, a centuries-old fig tree whose canopy spreads like a cathedral ceiling. Thin, silvery cables run from her branches into the BioLinguist Array, a shimmering wall of sensors that translates the plant's electrochemical signals into patterns of sound and light.

The council session begins. The city's planners have proposed building new floating gardens along the harbor to stabilize the shoreline. The plants respond in a chorus of soft pulses translated into slow, resonant speech by the Array. Aiya warns of a shift in the monsoon cycle and recommends a different planting pattern to avoid root rot in the coming decades. Her suggestion is not merely data; it's insight. The council takes note and adjusts plans accordingly.

In this world, we no longer plant crops without asking the crops themselves. Wheat fields in the northern plains hum with quiet electrical chatter, sending messages to human caretakers about soil moisture and nutrient needs. Rice paddies in the east "vote" on the best rotation patterns to keep pests at bay. The old word "farming" has been replaced with "collaboration."

Medicine has transformed, too. Patients recovering from trauma sometimes spend months in "Healing Groves," where selected plants chosen for their chemical profiles and responsive patterns are placed in proximity to the patient. The plant and patient engage in guided sessions where biofeedback devices let the human feel the subtle changes in the plant's own "mood." These sessions are part therapy, part conversation, each side learning to regulate the other's well-being.

But this partnership comes with new questions. Last year, a coalition of botanical networks petitioned the United Earth Assembly for representation, arguing that their intelligence, however different from ours, warrants political rights. Debates rage in every corner of society. Some see plants as our equals. Others insist that this is dangerous territory and that nature should never be recognized as a legal entity.

As I leave the council chamber, the wind shifts, carrying the scent of flowering jasmine from the rooftop terraces. It's subtle, but I feel as if the plants are aware of me leaving, their rustling a faint goodbye. In a world where we have finally learned to listen, even silence feels like a kind of language.

The vision of a future where humans and plants converse as partners might feel like pure imagination, but the roots of that idea run deep into both ancient traditions and modern scientific curiosity. Across continents and centuries, people have treated plants not simply as passive background to human life, but as conscious participants in the world. What's changing now is that science is beginning to ask questions that echo what Indigenous healers, shamans, and traditional farmers have been saying for generations.

Chapter 14 invites us to explore the concept of plant consciousness from both perspectives: how wisdom traditions have long spoken of plants as intelligent beings, and how modern plant science is testing the limits of what we think life can do.

Plant Awareness in Indigenous and Shamanic Traditions

In many Indigenous cultures, plants are regarded as more than biological organisms; they are kin, allies, and conscious partners. Among the Shipibo-Conibo people of the Amazon, for example, plants are believed to hold spirits that can guide human behavior and offer healing. In parts of North America, certain First Nations refer to trees as "standing people," acknowledging them as members of the same life community. Across Africa, sacred groves are treated as sentient guardians of the land, while in Asia, traditions like Shinto in Japan honor specific trees as repositories of divine presence.

These views share a common belief: plants have awareness, intention, and even agency, though expressed in ways unlike human or animal consciousness.

In traditional practice, human–plant communication often unfolds through structured rituals. Shamans, healers, and medicine people use song, chant, and offering to "speak" with plant spirits. In the Amazon, icaros or sacred songs are sung during healing ceremonies to call upon a plant's unique energy. In other cultures, offerings of tobacco, cornmeal, or flowers are made before harvesting, signaling respect and reciprocity. Some traditions also use plant-based divination, where leaves, seeds, or flowers are read for guidance in the same way others might interpret dreams or stars. Vision quests and dieta (periods of isolation and plant fasting) are used to enter altered states in which plant spirits are believed to speak directly to the seeker.

Intriguingly, many of these teachings have later found support in science. For example, certain medicinal uses known to Indigenous healers for centuries, such as willow bark for pain relief, were confirmed only after modern research identified active compounds, such as salicylic acid. This alignment raises questions: Did ancient people simply experiment endlessly, or were they learning through a kind of intuitive or relational process with plants?

Modern Plant Neurobiology Debates

Modern plant neurobiology explores whether plants can be considered intelligent without having brains or nervous systems. Researchers study plant signaling, memory, and problem-solving behaviors, such as anticipating the direction of light, recognizing related plants, or altering growth strategies to overcome obstacles. While these abilities don't mirror human intelligence, they suggest adaptive decision-making that goes beyond simple reflex.

Plants communicate internally using a network of electrical impulses, action potentials similar in nature (though not in structure) to nerve signals in animals. They also use hormonal pathways and complex chemical "languages" to warn neighbors of pests, coordinate flowering, or adjust metabolism. Root tips, in particular, are highly sensitive, detecting gravity, moisture, and even chemical traces of other organisms in the soil. Some experiments show root systems altering growth direction in response to these subtle cues, suggesting a form of "navigation."

Not all scientists agree on the term "plant intelligence." Critics warn against anthropomorphism, projecting human-like qualities onto plants, and argue that adaptive behavior does not equal consciousness. Supporters counter that intelligence should be redefined to include forms of problem-solving in non-animal life. The central challenge is methodological: how can we test for subjective experience in a being whose "thinking" may be entirely different from our own?

Famous Studies and Controversies

The Secret Life of Plants (1973)

In 1973, The Secret Life of Plants introduced millions of readers to the idea that plants might perceive, feel, and respond to human actions. The book described experiments where plants appeared to react to people's thoughts or distant events. While the cultural impact was immense (sparking public fascination with plant awareness), scientists criticized the work for a lack of rigorous controls.

Cleve Backster's Experiments

Polygraph expert Cleve Backster famously attached lie detectors to plant leaves, claiming they registered changes when humans thought about harming them. He reported even more astonishing results—plants reacting to events happening far away. These findings captured public imagination but failed to withstand replication in controlled laboratory conditions. Mainstream science rejected his methodology, yet the stories remain enduring symbols of curiosity about plant sentience.

Legacy and Influence

Despite skepticism, the work of Backster and the popularity of The Secret Life of Plants inspired generations of researchers to investigate plant signaling and sensory capacities with more precision. Some modern studies, though not as sensational, have found evidence that plants can register and respond to environmental changes far more dynamically than once thought. The debate over whether this amounts to "consciousness" continues, keeping alive one of the most intriguing questions in biology.

Toolkit

Here are simple ways to put this chapter into action in your own life:

1. Observation Without Assumption

Choose a single plant indoors, in your garden, or in the wild and spend ten minutes a day observing it without projecting feelings or expectations. Notice the smallest details: leaf movements, changes in posture, and color shifts. The goal is to witness without forcing interpretation, allowing awareness to deepen naturally.

2. Respect Ritual

Before tending or harvesting from a plant, create a small ritual of acknowledgment. This could be as simple as pausing for gratitude, offering a few drops of water, or speaking a quiet word of thanks. Indigenous traditions remind us that a relationship begins with respect, and the act of acknowledging a plant's life changes the way we interact with it.

3. Song or Sound Connection

Experiment with vocalizing or playing music near plants. Document any differences you notice over weeks, growth changes, leaf orientation, or simply how you feel in the plant's presence afterward. Some traditions believe song is a universal language between species; see what emerges for you.

4. Dream Dialogue

Before sleeping, hold a leaf, flower, or seed of your chosen plant in your hand or place it nearby. Set an intention to receive insight or guidance through your dreams. Upon waking, jot down any images, phrases, or feelings that come up, even if they seem unrelated. Over time, patterns may emerge.

5. Cross-Checking with Science

Take one plant-centered teaching from a traditional culture, such as a medicinal use or behavioral trait, and research modern scientific findings about it. Notice where they align, where they differ, and how both can inform your understanding. This exercise helps bridge ancient and modern perspectives without rejecting either.

6. Community Conversation

Ask friends, family, or community elders if they've ever felt a plant "communicate" with them. You may be surprised how many people have quiet stories they rarely share. Gathering these experiences reinforces the shared human connection to the plant world and expands your own sense of possibility.

Chapter 15: Communicating with the Green World

I still remember the first time I held a conversation with a tree.

It was early morning in New Kyoto, and the streets were quiet except for the soft hum of the vertical gardens woven into the sides of the buildings. I stood in the shade of a century-old sakura tree, its blossoms spilling pale pink light into the air. My neural interface was active, tuned to the Green Translation Network. This system connected every plant in the city to the biofeedback grid.

At first, I heard nothing. Just the low electrical murmur of the translator waiting for a signal. Then, gently, like a ripple on still water, a voice arrived. Not a human voice, more like a shifting arrangement of tones and fragrances, impossible to confuse with anything man-made. The translation rendered it in words: "You are early. The air is kind today."

I felt my throat tighten. The tree knew I was here. It had noticed me. I leaned against the smooth curve of its trunk, the translator relaying subtle shifts in its biochemistry, sugar flows, water movements, and the faint adjustments of leaf position in response to sunlight. Each was expressed as an image in my mind: a warm pool of gold, a stream of silver threads, a breath of wind.

The conversation wasn't small talk. The sakura asked for a soil pH adjustment last night's rain had brought too much acidity. It also requested more ladybugs in the nearby rose beds to help with a growing aphid problem.

But then the tone shifted, offering something unexpected. "The sparrows will leave early this year," it told me. "Prepare the northern planters for seed."

I didn't understand why it mattered until weeks later, when the sparrows migrated ahead of schedule, just as the sakura predicted. That early warning saved dozens of rooftop crops in the district.

Now, these exchanges are part of daily life. Farmers across the city consult their crops each morning as easily as checking the weather. In the Biospire towers, vertical forests hold council with human caretakers to design entire seasonal growth strategies. We don't just cultivate plants anymore; we collaborate.

And yet, I've come to realize the technology is only half the miracle. The other half is the willingness to listen, to let a green, rooted being set the rhythm.

The sakura never asks for much. Just sunlight, clean water, and for someone to stop, once in a while, and notice the air.

The sakura's voice faded back into the gentle hum of the Green Translation Network. Still, the feeling it left stayed with me as a quiet reminder that dialogue with plants doesn't always require advanced technology.

For centuries, humans have been speaking with the green world through stories, rituals, and silent observation. Before there were neural links and biofeedback grids, there were ceremonies, songs, and patient hours spent listening without expectation.

The idea that plants can "speak" isn't new; it's one of the oldest human understandings we have. It has been carried in myths, encoded in seasonal celebrations, whispered in dreams, and confirmed in subtle exchanges between hands, soil, and leaf.

What modern tools offer is simply another way to participate in something ancient. Whether through sacred ritual or scientific interface, the real key is the same: the willingness to listen.

Plant–Human Communication in Myth, Ritual, and Meditation

Across cultures, plants have appeared as voices of guidance, caution, or revelation. In Greek mythology, the sacred oak at Dodona was said to speak the will of Zeus through the rustling of its leaves, interpreted by priests. Celtic lore tells of yew trees guarding wisdom between worlds, their whispers carrying messages to those who sat beneath them at dusk. Indigenous legends from the Americas tell of corn, tobacco, or cedar offering counsel in times of need, often through dreams or visions.

In these traditions, plants are not inert backdrops to human life; they are messengers between realms. Their "speech" may come as sound, imagery, or sudden knowing. Symbolically, speaking plants remind us that wisdom is rooted in the natural world and available to those who approach with respect.

Long before modern agriculture, planting and harvest were framed as acts of relationship. Music, blessings, and offerings accompanied ceremonial planting days to the seeds. In some traditions, water, song, incense, or food was gifted to plants to open dialogue. Seasonal rites marked the blooming of certain flowers, the ripening of fruits, or the shedding of leaves not just as agricultural markers, but as conversations with the cycles themselves.

These rituals encode a form of reciprocity: humans care for plants, and plants in turn provide sustenance, medicine, or insight. The act of making an offering, however small, signals that communication is a two-way exchange.

For many, the most direct form of plant communication comes through stillness. Sitting with a single tree, flower, or sprout and observing without an agenda can bring an uncanny sense of presence. Guided visualizations, where one imagines meeting and speaking with a plant spirit, can deepen this bond. Mindfulness practices, simply noting color, scent, texture, and subtle changes, can begin to feel like a conversation in slow motion. Over time, a person may notice patterns: a certain plant seeming to "perk up" when approached, or a shift in mood when in its company.

Practices for Listening to Plants

Experienced herbalists often describe a moment when a plant "calls" to them, either visually standing out among many or presenting itself in dreams. They may "hear" which part to harvest or how to prepare it, relying on both botanical knowledge and intuition. Some remedies in herbal traditions were reportedly discovered not through trial and error, but through this felt communication.

In permaculture, one of the first principles is observation before action. This means spending days, weeks, or even seasons simply watching the way plants grow, how they interact with sunlight, wind, and each other, and how they respond to changes. The gardener notes subtle signals, leaf curl, posture shifts, or scent changes that point to needs or imbalances. This "listening" through observation often prevents mistakes and leads to more harmonious cultivation.

Plant communication isn't purely visual. The sound of leaves under different weather conditions can reveal changes in hydration. Touch may detect tension or slack in stems, while scent shifts can signal flowering, fruiting, or stress. Some practitioners speak of sensing a change in "energy" when near certain plants, a warmth, tingling, or lightness that is difficult to quantify but deeply felt.

Anecdotal Accounts of Plant Guidance

Gardeners sometimes speak of being drawn to a plant without knowing why, only to discover later that it offered exactly the remedy they needed. Others tell of plants that seem to flourish only under certain caretakers, as though responding to a particular person's presence. Some even note uncanny timing, where a plant's flowering or seeding coincides with a meaningful event in their life.

Indigenous healers recount receiving instructions from plants during dreams or ceremonies, guidance not just for physical healing, but for resolving disputes, restoring balance, or teaching moral lessons. In these traditions, there are protocols for receiving plant wisdom, such as fasting, offering prayers, or conducting specific rites before engaging in dialogue.

In recent years, musicians and researchers have begun using biofeedback devices to translate plant electrical activity into sound. The result is hauntingly melodic "music" generated by the plant's own rhythms. Others experiment with spoken word, vibration, or frequency-based communication, sometimes claiming noticeable changes in plant growth or vitality. While skeptics caution that these results are often anecdotal, the experiments continue to spark curiosity and perhaps remind us that communication may take many forms beyond human language.

Toolkit

Here are simple ways to put this chapter into action in your own life:

1. The Single-Plant Sit

Choose one plant and sit quietly with it for at least 10 minutes each day for a week. Observe without touching at first. Note changes in posture, leaf angle, color, and any subtle sensory impressions (scent shifts, air temperature, sound of leaves). By the end of the week, write down any patterns or feelings that emerged.

2. Offering Ritual

Select a way to make a small offering to a plant or garden. This could be a pinch of herbs, a song, a bowl of fresh water, or a silent blessing. Perform the offering mindfully, acknowledging the plant's presence. Notice if this ritual changes your sense of connection over time.

3. Dream Connection Exercise

Before bed, focus your attention on a plant you wish to "speak" with. This can be a living plant nearby, a photo, or even a species you've read about. Set the intention to receive guidance in dreams. Keep a notebook by your bed and record any dream fragments or waking impressions the next morning. Look for recurring symbols or themes.

4. Cross-Sensory CheckIn

Spend five minutes using each sense, one at a time, to "listen" to a plant.
– Sight: observe shape, movement, and light reflection.
– Smell: note scent strength, changes.
– Touch: gently feel texture, tension, or softness.
– Hearing: listen for rustling, creaks, or subtle sound shifts.
– Intuition: notice any "gut feelings" or emotional impressions.

5. Permaculture Observation Journal

If you have access to a garden, farm, or wild patch, commit to observing the same area at the same time each day for at least two weeks. Record changes in plant growth, patterns of insect or animal visitors, and interactions between plants. This builds an awareness of seasonal rhythms and plant "signals" over time.

6. Plant Sound Experiment

If possible, use a basic plant music device or an online plant biofeedback tool. Spend time with the plant while the device is active and note how the sound changes with your presence, voice, or touch. Treat this as exploration, not proof. Let the experience guide your reflection.

7. Shared Story Circle

Gather with friends, family, or an online group and share one personal story of feeling connected to a plant. Listen for similarities in language, imagery, or emotional tone. This practice reinforces that plant–human dialogue is a shared and ancient part of human experience.

Chapter 16: The Human Biofield and Plant Energy

The first thing you notice when stepping into the Biofield Garden is the air. It's not just fresher, it feels charged, alive, like the moment before a summer storm when every hair on your arms stands to attention. I ease myself into a low wooden bench, careful not to disturb the moss spreading thick and green at my feet.

Across the path, a man with a healing incision along his side reclines beside a cluster of luminous blue ferns. Above each plant, faint holographic threads of light ripple outward, merging in midair with a shifting aurora that hovers around him. I've read about this: the quantum biofield sensors translate the plants' subtle electromagnetic output into visible waves, revealing the constant, invisible conversation between species.

A nurse moves slowly between patients, holding a slim tablet. She kneels beside a woman in a wheelchair, adjusting the angle of a potted ficus. On the display hovering above them, the woman's pale, fractured aura begins to fill in with soft greens, matching the plant's steady glow. "Better alignment now," the nurse says quietly, almost like she's speaking to both patient and plant.

I look to my left. A child with a shaved head from recent chemotherapy is cupping a small jasmine plant in his hands. The hologram above them flares brilliant white and gold so bright that nearby patients turn their heads. He laughs, and the jasmine seems to tremble with him, its petals unfolding just a fraction more. The nurse doesn't need to explain. The connection is visible to everyone.

In this place, healing is no longer a solitary battle waged inside a body. It's a partnership, an exchange. The plants take in our carbon dioxide and fractured frequencies, returning oxygen, biophotons, and an unquantifiable calm. Every pairing is intentional: lavender for pain, lemon balm for sleep, spiraling ficus for trauma recovery. Years of biofield data have proven what ancient traditions whispered: plants are not passive décor. They are active healers.

By the time I stand to leave, my own holographic outline has shifted. The jagged edges smoothed, the dim spots now softly glowing. The fern beside me sways in still air, as if waving goodbye. I can't tell if the change is measurable in molecules or only in me, but I know I am walking out lighter than I came in.

What you just experienced in that imagined Biofield Garden is not entirely fantasy. Many of the concepts, such as plants interacting with our energy fields, the measurable exchange of subtle electromagnetic signals, and the intentional design of healing environments, are already being studied, practiced, and, in some cases, implemented in the real world. While our technology today is not yet as visually striking as the holographic auras of 2075, both ancient traditions and modern science point toward the same truth: humans and plants share more than oxygen and carbon dioxide. We share energy. In this chapter, we'll explore the human biofield, how plants might interact with it, and the scientific, cultural, and technological bridges being built to understand this invisible relationship.

Understanding the Human Biofield

The human biofield is described as a dynamic, constantly shifting electromagnetic field that surrounds and permeates the body. In energy medicine, it is believed to be a blueprint for physical, emotional, and even spiritual health. Ancient cultures have long had names for this life force: prana in India, qi in China, and ka in ancient Egypt. While terminology differs, the underlying concept remains consistent: health is linked to the balanced flow of this subtle energy.

In modern complementary and alternative medicine, the term "biofield" serves as a bridge between ancient wisdom and contemporary research. It is recognized not only in spiritual circles but also in academic settings, where integrative healthcare approaches are explored.

Scientific and Theoretical Foundations

From a scientific perspective, the body produces measurable bioelectromagnetic fields, most prominently from the heart and brain. These fields are detected through medical devices like electrocardiograms

(ECGs) and electroencephalograms (EEGs). Beyond these, researchers have found evidence of extremely weak light emissions from all living cells, called "biophotons." These faint flashes of light are thought to play a role in cellular communication and possibly in maintaining overall coherence in the body.

Measuring these subtle energies is challenging. Current instruments are limited in sensitivity, and the debate continues as to whether the biofield is purely physical, purely energetic, or a combination of both.

The Biofield's Role in Health and Healing

Energy medicine practitioners often describe the biofield as a reflection of a person's internal state. A smooth, strong field is associated with vitality, while a disrupted or "collapsed" field may be linked to illness, fatigue, or emotional distress.

Practices such as Reiki, qigong, and therapeutic touch work directly with the biofield, using either hands-on or hands-off techniques to restore balance. While scientific validation is still developing, many people report improved well-being, reduced pain, and enhanced relaxation after such interventions.

Plant Interaction with Human Energy Fields

Many cultural traditions describe plants as possessing energetic fields, often called plant auras. Indigenous knowledge systems often hold that certain plants can absorb heavy or negative energies, transforming them into lighter, more harmonious vibrations. Modern observational reports note that some plants seem to lose vitality in the presence of prolonged human stress. In contrast, others appear to flourish with loving attention. The connection between plants and human energy is not just poetic; it can be physically felt. Many people describe an almost tangible sense of renewal when they spend time in forests, gardens, or even with a single potted plant. Placing hands on a tree trunk or digging into the soil is believed to ground and harmonize human energy, a concept echoed in the Japanese practice of shinrin-yoku or forest bathing. In this framework, plants do not simply filter our air; they also filter our energy.

From feng shui to Vastu Shastra, the arrangement of plants within a space has long been tied to energy flow. In meditation rooms, plants are chosen not only for their appearance but also for their calming influence. Hospitals and healing centers often use greenery to speed recovery, a practice now supported by studies showing faster healing rates, reduced stress hormones, and improved mood in plant-rich environments.

Kirlian Photography, Aura Work, and Modern Interpretations
Kirlian photography, developed in the 1930s by Semyon and Valentina Kirlian, captures images of corona discharges, tiny luminous halos around objects, when they are placed on photographic plates under high voltage. Early experiments showed distinct patterns around leaves, fingertips, and other organic matter. Enthusiasts claimed these images captured the aura or life force, while skeptics pointed out that the patterns could be explained by moisture, pressure, and conductivity.

Some energy healers claim they can see or sense the auras of plants directly, noting differences in color, shape, and brightness as indicators of health or mood. A robust, vibrant aura may be interpreted as a sign of a thriving plant, while dull or fragmented patterns might suggest stress or disease. These interpretations remain controversial, with critics questioning their subjectivity and supporters arguing for more open-minded exploration.

Today, digital aura imaging systems and biofeedback devices are making it easier to experiment with human–plant energy interactions. Some tools translate plant electrical signals into sound, creating music that changes in response to touch, proximity, or even emotional states. Others attempt to map subtle energy patterns in color or light. While mainstream science remains cautious, the merging of traditional concepts and modern technology may one day provide a clearer picture of how human and plant energies interact and perhaps confirm that the "Biofield Gardens" of the future are not as far away as they seem.

Toolkit
Here are simple ways to put this chapter into action in your own life:

1. Hands–Tree Energy Exchange

• Find a healthy tree in a quiet area. Stand with your back against the trunk or place your palms directly on the bark.

• Close your eyes and breathe slowly. Imagine your breath drawing in strength from the tree as you exhale any stress or heaviness.

• Notice changes in body sensation—temperature, tingling, relaxation—without judgment.

• Repeat for 5–10 minutes, and record any observations afterward.

2. Plant Biofield Sensing

• Choose a living plant (houseplant, herb, or garden plant) that you can interact with daily.

• Stand or sit about one foot away, palms facing it. Slowly move your hands closer until you feel subtle changes—such as warmth, coolness, tingling, or resistance.

• Keep a journal of sensations and note any patterns over several days or weeks.

3. Biophoton Meditation

• In a dimly lit room, place your hands in front of your closed eyes. Visualize a soft glow radiating from your palms into the space between you and a plant.

• Picture the plant responding with its own light, meeting yours in the middle.

• This is a mental exercise designed to increase focus on subtle energy connection, whether or not light is physically visible.

4. Healing Space Arrangement

• Select one area in your home where you spend time relaxing or meditating.

• Add plants with calming qualities (such as peace lilies, snake plants, or lavender).

• Position them so they create a natural visual flow—avoid sharp, blocking lines.

• Spend time daily in this space and track any shifts in mood, focus, or physical well-being.

5. Forest Bathing Biofield Reset

• Visit a wooded area or botanical garden. Turn off your phone and remove distractions.

• Walk slowly, pausing often to touch leaves, smell flowers, or simply stand still and listen.

• Visualize your personal energy field expanding to blend with the living field of the plants around you.

• At the end, express gratitude silently or aloud before leaving.

6. Kirlian Photography Exploration

• Research a local practitioner or workshop offering Kirlian photography or digital aura imaging.

• Compare your aura image before and after spending time with plants.

• Keep in mind scientific debates around these tools—focus on personal insight rather than absolute proof.

7. Sound–Plant Interaction

• If available, use a plant music device (such as one that translates electrical signals into sound).

• Speak gently, play soft music, or even meditate aloud near the plant and observe sound changes.

• Reflect on whether different emotional states create noticeable differences in the plant's musical "expression."

Chapter 17: Plants as Healers of Spirit

I stepped through the arcology's security gates and into the Healing Enclave, leaving behind the hum of drones, the glare of glass towers, and the metallic tang of the city's recycled air.

The moment the doors sealed behind me, the world changed. Warm, living air flowed over my skin, heavy with the scent of sage, cedar, and palo santo. Above me, a canopy of bioluminescent leaves glowed faintly, their light pulsing in a slow, soothing rhythm. The entire greenhouse seemed to breathe with me.

Kael, the enclave's healer, met me barefoot, her hair tied back in a simple knot. "You've been carrying too much," she said, not as a guess, but as if she could see the heaviness draped over my shoulders. She led me down a path lined with aromatic plants, their leaves brushing against my arms like curious hands.

We stopped beside a circular altar where cedar boughs lay in symmetrical spirals, each branch tagged with a tiny holographic marker showing its lineage grown from seeds preserved since the 21st century, long before mass extinction reshaped the world. The air shimmered faintly around them, not with heat, but with energy.

Kael tapped a small device on her wrist. A resonance scanner unfolded from its housing, projecting a lattice of light over my body. It mapped my biofield in three dimensions, revealing streaks of distortion, places where my energy was tangled and dim.

She selected a bundle of sage from a hydroponic cradle. "This one's tuned to your field," she explained. "It'll meet you where you are, then lead you where you need to be."

She lit the sage with a spark from her palm-sized plasma torch, and the smoke rose in curling waves. It wasn't just scent; it carried a subtle vibration that I could feel, a warm humming under my skin. As she circled me, the biofield distortions on the scanner began to fade, replaced by brighter, cleaner streams of light.

When she finished, Kael guided me to the palo santo grove at the heart of the greenhouse. The trees were small but ancient in spirit, their DNA traced to forests in Ecuador that had been destroyed generations ago. She placed my palm against one trunk. A cool, electric pulse traveled up my arm, and for a moment, I felt the memory of rainfall, birdsong, and sunlit air, the tree's own archive of life.

"You're in conversation with them now," Kael said softly. "They've been waiting to help you remember who you are."

When I left the enclave, the city's noise seemed smaller, its grip weaker. The fragrance of cedar clung to my skin like a shield, and somewhere deep inside, I felt rooted again. In a world that had tried to outgrow nature, the plants still remembered us and still spoke, if we were willing to listen.

The greenhouse's doors whispered shut behind me, and for a long moment, I stood in the corridor, feeling the cedar-scented air still clinging to my clothes. The city beyond was as it had always been: fast, loud, and impatient, but something in me had shifted. I understood then what ancient healers, shamans, and medicine keepers had always known: plants do more than heal the body. They reach into the unseen spaces within us, clearing away what we no longer need and anchoring us back into the web of life.

This truth has been carried forward through countless traditions, sometimes quietly, sometimes in the smoke of sacred fire, sometimes in the petals laid on graves or the green garlands woven into wedding crowns. Whether in ceremony, in private meditation, or in futuristic healing enclaves, plants continue to serve as allies to the human spirit. In this chapter, we will explore how cultures past and present, and perhaps our future selves, have worked with the spiritual medicine of plants.

For countless generations, sage has been revered as both a protector and a purifier. In North America, white sage (Salvia apiana) holds a sacred place among Indigenous nations, often used in smudging rituals to cleanse people, spaces, and ceremonial tools of unwanted energies. The smoke is believed to carry prayers upward, creating a bridge between the human and spirit worlds.

Not all sage is the same. Garden sage (Salvia officinalis), common in European folk traditions, has been used in home blessings and kitchen magic. Other regional species, such as desert purple sage, carry their own distinct fragrance, symbolism, and ceremonial uses.

The sensory experience of smoke cleansing is part of its power: the earthy aroma, the curling tendrils of smoke, the deliberate, mindful motions of waving the bundle. This sensory ritual signals to the mind and spirit that a threshold has been crossed, that one is entering a protected, sacred space.

Cedar has been called "the tree of life" in many cultures, valued not only for its durability but for its spiritual strength. Among numerous North American Indigenous communities, cedar branches are burned in smudging ceremonies for protection, grounding, and blessing. It is said to drive away negative influences and invite in peaceful, stabilizing energy. The symbolism of cedar is deeply tied to longevity and resilience. In nature, cedar trees can live for centuries, weathering storms and harsh climates. In spiritual practice, cedar is thought to pass this endurance to those who work with it. Some traditions place cedar boughs on the floor during sweat lodge ceremonies, while others use cedar-infused baths to restore balance after illness or emotional distress.

In South America, particularly in Peru and Ecuador, palo santo or "holy wood" is burned for purification, protection, and the calling in of blessings. Its sweet, resinous smoke is said to lift the spirit, dispel melancholy, and attract positive energy.

Traditionally, palo santo is harvested only from naturally fallen branches, ensuring that the tree is respected and allowed to complete its life cycle. Modern demand, however, has raised sustainability concerns, making ethical sourcing essential for those who wish to honor the plant's spirit. When used in a ceremony, palo santo is often combined with intention-setting. The aromatic experience works in concert with the ceremonial act, creating a multisensory moment of connection between the human heart and the wider living world.

Cultural Ceremonies and Symbolic Plant Allies

Across cultures, plants have been central to rituals marking the turning of the seasons and the cycles of planting and harvest. In Europe, Lammas marked the first wheat harvest with offerings of bread and grain. In parts of Asia, rice festivals honored the spirit of the crop with dances and feasts.

Solstice and equinox celebrations often involve plant offerings, flowers, fruits, or branches as symbols of gratitude and renewal. These ceremonies encode the understanding that human life is inseparable from plant life, and that honoring this bond ensures abundance and balance.

Plants in Life Transitions

From birth to death, plants have been woven into the milestones of human life. In weddings, roses or orange blossoms symbolize love and prosperity. At funerals, lilies, marigolds, or evergreen boughs carry messages of remembrance and eternal life.

Coming-of-age ceremonies in many cultures include plants as protective talismans or as initiatory symbols, herbs tucked into clothing, garlands worn on the head, or sacred seeds planted to mark the passage into a new stage of life.

These rituals not only honor the individual but also reaffirm their place within the community and the natural world.

Plants as Symbols in Story and Song

Oral traditions worldwide are filled with stories where plants embody virtues, warnings, or lessons. A tree might shelter a hero, a flower might signal the presence of a spirit, or a vine might represent unbreakable bonds.

Songs and chants invoking plant allies often serve dual purposes: they pass down botanical knowledge and carry the vibrational qualities believed to strengthen the connection between humans and plants.

Through these stories and melodies, cultural plant wisdom is preserved, sometimes for centuries, without ever being written down.

Toolkit

Here are simple ways to put this chapter into action in your own life:

1. Preparing Your Space for Plant Rituals
• Choose a quiet location where you won't be disturbed.
• Open a window or door if you're working with smoke to allow energy to flow and prevent stagnation.

• Place a natural surface, wood, stone, or earth between yourself and the plant material when possible.

2. Smoke Cleansing with Sage, Cedar, or Palo Santo

• Light the tip until it smolders, producing smoke without a large flame.

• Fan the smoke gently with a feather or your hand, guiding it toward your body, an object, or a space.

• While moving the smoke, repeat an intention or prayer, such as "I release what no longer serves me and invite in peace."

• Extinguish by pressing the burning end into sand or earth, never in water unless necessary.

3. Non-Smoke Alternatives

• Create a plant-infused water spray using a few drops of essential oil in purified water.

• Place dried herbs in a bowl and stir with your hand while focusing on your intention.

• Work with fresh plant material in a vase or bundle, allowing its scent and presence to influence the space.

4. Seasonal and Ceremony Integration

• Incorporate local plants into solstice or equinox rituals to align with your environment's cycles.

• Use flowers or herbs significant to your culture during life events, weddings, naming ceremonies, or memorials.

• Keep a journal of the plants used in each ceremony and how they influenced the mood or energy.

5. Meditative Connection Practices

• Sit with a plant in silence for 5–10 minutes daily, observing its form, scent, and energy.

• Imagine a thread of light connecting your heart to the plant's center.

• Ask internally for any guidance or message, and listen without forcing an answer.

6. Ethical and Sustainable Sourcing

• When working with plants like white sage or palo santo, ensure they are ethically harvested from sustainable sources.

• Grow your own ceremonial herbs whenever possible to deepen your personal connection.

• Learn and honor the cultural origins of the plant traditions you practice.

7. Combining Plants with Sound and Song

• Use a drum, rattle, or singing bowl while holding or standing near a plant ally.

• Sing or hum melodies associated with the plant, or create your own in the moment.

• Pay attention to how the plant's scent, texture, or presence changes the feeling of the song.

8. Reflection and Integration

• After working with a plant ceremonially, sit quietly and notice any physical sensations, emotional shifts, or mental clarity.

• Write down insights, dreams, or synchronicities that occur in the following days.

• Thank the plant (verbally or silently) before ending your session.

Chapter 18: Plant Companionship and Emotional Resonance

I didn't move here for the plants. I moved here because the rent was cheap and the building was close to my office, though "close" meant little when I was working sixteen-hour days in front of a holo-screen.

The leasing agent called it a "bio-symphony apartment," as if that made sense. I pictured a trendy marketing gimmick: green walls, fake nature sounds, maybe a vase of overpriced flowers. But the first night I walked in, the air felt different. It wasn't just oxygen-rich. It was alive.

The entire far wall of the studio pulsed with greenery, ferns, vines, and strange blooms that seemed to tilt their faces toward me like curious animals. The leaves shimmered slightly in the dim light, and I told myself it was just my eyes adjusting.

The building's AI, a soft-voiced program named Lira, explained the "bonding process." These plants, she said, had been genetically tuned to respond to human biofields, our subtle electromagnetic signatures.

"They'll adjust to your emotional patterns over time," she said, as if I'd just adopted a particularly quiet pet.

At first, nothing much happened. I worked. I slept. I forgot to water them, and Lira reminded me with polite chimes. But on the fourth evening, I came home after a brutal project deadline and noticed a faint change in the air, an herbal sweetness, sharp at first, then warm, almost like cinnamon. Lira informed me the wall had released a custom blend of terpenes "to soothe heightened stress response." I laughed, but it worked. My shoulders unclenched.

By the second month, I was paying attention. When I entered in the mornings, the leaves tilted slightly toward me, tracking my movement like sunflowers following the sun. On days I felt good, the foliage deepened to a richer green; on the days I dragged myself home in a fog, the wall was softer in color, as if it was resting with me.

One night, I asked Lira to play the plants' "song." Hidden sensors converted their electrical activity into sound, and the apartment filled with slow, resonant tones, a kind of living music that vibrated in my chest. It didn't feel like the plants were performing; it felt like they were speaking.

The strangest thing happened when I left for a week-long work trip. I expected to feel relief at escaping the city, but halfway through, I missed them. I missed the way they seemed to notice me.

When I finally returned, the apartment was quiet. The leaves didn't move right away. I set down my bag, stepped closer, and rested my palm against a broad, cool leaf. Slowly, the wall seemed to wake tiny movements, a deepening of green, that warm, familiar fragrance blooming into the air.

That's when I realized: I hadn't been living alone for months.

While not every home has a bio-responsive living wall, the essence of that experience, genuine companionship with plants, is available to all of us. It appears in ancient traditions, modern therapy programs, and everyday relationships between people and the green life they nurture. The simple act of tending to plants is more than a hobby; it is a form of emotional connection that can ground, heal, and transform us.

Which brings us to our next exploration: how plant companionship creates emotional resonance and supports human well-being in both ordinary and extraordinary ways.

The Comfort of Tending to Plants as a Form of Therapy

Horticultural therapy is the intentional use of plant-related activities to improve mental, physical, and social well-being. It is not new; its roots reach back centuries, from monastery gardens used to soothe the sick to the moral treatment movement of the 19th century, where hospital gardens were designed to lift the spirits of patients. In modern times, horticultural therapy is used in hospitals, rehabilitation centers, elder care homes, and mental health programs. Participants engage in tasks like planting seeds, pruning, or arranging flowers. Documented benefits include reduced anxiety, improved mood, enhanced motor skills, and greater social interaction.

Caring for plants offers more than visual beauty; it can help regulate emotions. A consistent plant care routine provides structure and purpose, anchoring the mind during turbulent periods. Repetitive, mindful gardening actions, such as watering, repotting, and trimming, invite a meditative state, lowering stress hormones and calming the nervous system. Many people describe their plants as silent confidants, "listening" without judgment during moments of frustration, sadness, or reflection. The simple act of tending to a living thing and seeing it thrive can be deeply therapeutic. Watching a sprout emerge from soil or a bud open into bloom fosters feelings of accomplishment and hope. This nurturing relationship can extend inward, encouraging self-compassion. Much like pet therapy, plant care invites emotional warmth, responsibility, and an ongoing exchange of care that benefits both the plant and the person.

Houseplants as Energetic Companions

Even without active interaction, plants can influence the atmosphere of a space. Their colors, shapes, and subtle scent can create feelings of calm, focus, or vitality. People often describe a room with plants as "more alive." In some cases, owners name their plants, speak to them, and regard them as friends and companions that witness daily life without judgment or demand.

Over time, many plant owners notice differences in how various species "feel." Some plants seem serene and grounding; others feel bright and uplifting. Anecdotal accounts suggest that plants may respond differently to individual caretakers, as if showing preferences or moods. The bonds formed through years of care, especially for long-lived species, can carry significant emotional weight, akin to a friendship.

During periods of isolation, such as lockdowns or long-term illness, houseplants have provided crucial companionship. People have turned to their plants for comfort, stability, and the reassurance of daily interaction. Online communities of "plant parents" have flourished, where people share stories, tips, and even photos of their plants as if introducing beloved family members.

Gardening as a Spiritual Practice

Gardening can be a powerful form of moving meditation. Each action, digging, planting, watering, draws attention to breath, movement, and sensation. Focusing intently on a single plant or task anchors the mind in the present moment, creating a refuge from distraction and worry.

The natural cycles of planting, growth, harvest, and decay mirror human life stages. Planting seeds can symbolize hope or intention-setting, while pruning or composting represents release and transformation. These cycles remind us of life's impermanence and renewal, deepening our connection to the natural order.

For many, gardens are sanctuaries, places of healing, prayer, or quiet reflection. Some incorporate altars, spiritual symbols, or offerings among the plants, blending aesthetic beauty with sacred purpose. Gardens can also serve as spaces of grief, where planting or tending becomes a ritual of remembrance and restoration.

Toolkit

Here are simple ways to put this chapter into action in your own life:

1. Create a Plant Care Ritual

Choose a specific time of day (morning coffee, lunch break, or evening wind-down) to tend to your plants. Use this time to water, prune, or simply observe them. Treat it as a mindful practice, moving slowly and noticing each detail: the scent of the soil, the texture of the leaves, the light on the stems.

2. Plant a Symbolic Seed

Select a plant to represent a personal goal, hope, or healing process. Write your intention on a small slip of paper, bury it in the soil, and tend the plant as a reminder of your commitment. Let the plant's growth mirror your own progress.

3. Emotional CheckIn with Your Plants

Stand or sit quietly near your plant and take a few deep breaths. Notice your emotional state. Speak your thoughts out loud or silently. Observe the plant without trying to change it, allowing its stillness to influence your own.

4. Energetic Atmosphere Experiment

Place a plant in a room where you spend a lot of time, such as your workspace or bedroom. Over a week, note any shifts in your mood, focus, or energy. Try moving the plant to another room and observe if the atmosphere changes.

5. Companion Plant Journal

Give each of your plants a name and keep a journal documenting their growth, changes, and care. Include your own reflections on how you were feeling during each interaction, what you noticed about the plant, and any symbolic connections you sense.

6. Create a Sacred Green Space

Designate a corner of your home or yard for plant companionship. Arrange plants in a way that feels harmonious to you. Add objects that hold meaning, such as stones, candles, or photos, to make it a personal sanctuary for reflection or meditation.

7. Isolation Connection Ritual

If you live alone or are experiencing isolation, choose one plant as your "anchor" plant. Make a habit of speaking to it daily, sharing your thoughts, reading to it, or simply sitting beside it. This simple interaction can create a comforting routine and sense of connection.

8. Forest Bathing at Home

If you cannot access a park or forest, create a "mini-forest" by grouping several plants together in one space. Spend at least 15 minutes a day sitting among them, focusing on your breath and the subtle life energy around you.

9. Healing Through Gifting

Propagate a plant and gift it to someone going through a challenging time. Include a note explaining the meaning of the plant and how to care for it, turning your own connection into a ripple of healing.

10. Garden Gratitude Practice

If you have an outdoor garden, end each gardening session by placing your hands on the soil and expressing gratitude, whether aloud or silently, for the life it supports and the nourishment it provides you in return.

Chapter 19: Plants in Dreams, Visions, and Symbolism

I had read the reports, but nothing prepared me for the first moment inside the Neural Garden.

The technicians fitted the crown over my head, its metal filaments cool against my scalp. I blinked once, and the white lab dissolved around me. Suddenly, I was standing in a place that seemed to hover between dream and waking, a vast expanse bathed in soft, pearlescent light. The air shimmered, and in it floated flowers unlike any I'd ever seen: luminous lotuses, each one breathing with a slow rhythm, their petals glowing as though lit from within.

I reached out to touch one. The instant my fingers brushed its surface, it unfolded like a living scroll. Light spilled from its center, and within that light was a memory, not one I had chosen, but one it had chosen for me. I was nine years old again, in my grandmother's kitchen. The scent of jasmine tea curled through the air, and her hands were wrapped around mine, small and warm. She leaned in close, her voice soft but certain.

"Listen to the roots," she said. I had never heard those words before, not in this life, not in this memory. But the lotus released them into me as if I had always carried them.

When the vision faded, another lotus floated toward me. This one pulsed faster, as though impatient. I touched it, and the world around me shifted again. I stood beneath the massive canopy of an ancient oak. Its trunk was wide enough to hide a house, its bark furrowed with centuries of stories. Two paths stretched before me: one a golden trail lit with sunlight, the other dark, lined with whispering shadows. My chest ached with the weight of a choice I didn't yet understand. I tried to step forward, but the scene collapsed, the petals closing like a secret not ready to be told.

Lotus after lotus drifted past, each carrying fragments, glimpses of a life I had lived, might live, or perhaps had only dreamed. But all of them seemed connected, as though part of a greater map. Somewhere beneath my feet or perhaps beyond my mind, there was a hum. Not mechanical, but alive, like the deep vibration you feel when leaning against the trunk of an old tree.

And then I heard it. A voice, neither male nor female, soft yet carrying the weight of centuries.

"These are not dreams," it said. "We are the roots. We are your memory and your becoming. And we have waited for you to listen."

The petals began to close around me, the light folding in, and I felt myself returning to the lab. The crown lifted from my head, and the technicians exchanged quick glances. One of them leaned forward.

"Well?" he asked. "What did you see?"

I hesitated, the hum still echoing in my bones. "Not what I saw," I said. "It's what I met?"

The hum of the Neural Garden stayed with me long after I left the lab, vibrating somewhere deep in my chest. It wasn't just a memory; it was a message, an ancient one that seemed to speak in the language of roots and petals. I began to wonder if the lotuses, the oak, and the other visions weren't random creations of the system, but living symbols trying to guide me.

This is the timeless role of plants in dreams and visions. Across centuries and cultures, they have been our messengers, our teachers, and our mirrors, appearing in the sleeping mind with a purpose, carrying meaning both universal and deeply personal. What I experienced in the Neural Garden may have been built with technology. Still, the symbols themselves were far older than the machine that delivered them.

To understand why plants appear in our inner landscapes, whether in a dream, a vision, or a moment of deep meditation, we must explore the archetypes, sacred stories, and symbolic roles they have held in cultures across the world.

Archetypes and Plant Symbolism Across Cultures

The lotus is one of the most enduring and widely recognized plant symbols in human history. In Hinduism and Buddhism, it represents purity, enlightenment, and spiritual awakening. Its botanical qualities, emerging pristine from murky waters, reinforce its meaning as a symbol of transcending difficulties to achieve higher consciousness. Modern adaptations often use the lotus as a metaphor for personal growth, resilience, and the journey toward self-realization.

In Celtic and Norse traditions, the oak symbolizes strength, endurance, and wisdom. It played a sacred role in Druidic rituals and seasonal celebrations, often associated with the turning of the year. The oak's long lifespan serves as a metaphor for stability and steadfastness, its deep roots anchoring it through centuries of storms and change.

The yew has deep associations with death, rebirth, and immortality in European folklore. Frequently planted in churchyards and cemeteries, it was seen as both a protective symbol and a bridge between worlds. Its evergreen nature represents eternal life, while its ability to regenerate from seemingly dead wood embodies renewal and transformation.

Plant Spirits in Folklore and Religious Texts

Many cultures tell stories of plants that house protective spirits, capable of warding off evil or guiding the lost. Native American and African legends speak of shape-shifting plant beings, sometimes appearing as humans to deliver warnings or blessings. In some traditions, trees serve as dwellings for ancestral spirits, making them living vessels of memory and wisdom.

Plants appear prominently in sacred texts. In Judeo-Christian tradition, the Tree of Life and the Tree of Knowledge stand at the heart of humanity's origin story. In Buddhism, the bodhi tree is the site of the Buddha's enlightenment. Olive branches, in many traditions, symbolize peace, divine covenant, and reconciliation. These plants serve as both literal and symbolic anchors for faith.

Across cultures, plants have been seen as intermediaries between humans and the divine. Legends describe flowers or leaves used to carry prayers to higher realms. In some traditions, dreams or visions of specific plants are interpreted as direct messages, offering guidance, warnings, or blessings.

Dream Interpretation Involving Plants

Dreams of gardens, blossoming trees, or lush landscapes often represent creativity, prosperity, and untapped potential. Seasonal imagery, such as spring buds or autumn leaves, can reflect life stages or emotional cycles. Planting seeds in a dream is frequently interpreted as setting intentions or beginning a new chapter in life.

Withering or dying plants in dreams may signal emotional depletion, burnout, or unresolved grief. Conversely, dreams of receiving herbal remedies or plant-based medicine often represent healing, guidance, and spiritual support. Poisonous plants in dreams can symbolize hidden threats or situations that appear appealing but are dangerous at their core. Some dreams and visions involve plants that take on human-like qualities, speaking or interacting with the dreamer directly. Shamans and dreamworkers in multiple cultures use such plant visions for insight, healing, or divination. Intriguingly, there are many accounts of similar plant encounters in different parts of the world with no cultural contact, suggesting a universal language of plant symbolism embedded in the human psyche.

Toolkit

Here are simple ways to put this chapter into action in your own life:

1. Keep a Plant Dream Journal

Record dreams or visions involving plants as soon as you wake. Note the plant's appearance, condition, location, and any emotional impressions. Even small details like the color of a leaf or the feel of the soil can hold meaning. Over time, patterns will emerge.

2. Research the Plant's Cultural and Spiritual Meaning

Once you've identified a plant from your dream, look up its symbolic roles across cultures. Notice where meanings align and where they differ. This can reveal whether your dream's message is rooted in universal archetypes or personal associations.

3. Connect with the Plant in Waking Life

If possible, spend time with the plant you dreamed about. Visit it in nature, grow it at home, or work with it in dried or essential oil form. Physical connection often deepens understanding and helps translate symbolic messages into actionable insights.

4. Use Visualization to Continue the Dialogue

In meditation, return to the dream scene. Imagine speaking with the plant or walking among the landscape it appeared in. Ask it questions, and be open to receiving symbolic answers through images, sensations, or sudden thoughts.

5. Differentiate Between Symbolic and Literal Messages

Some plant dreams are purely symbolic, like a lotus rising from muddy water to signify spiritual growth, while others may be nudging you toward practical changes, such as adopting herbal remedies or adjusting your environment. Learn to sense the difference.

6. Honor the Plant's Message with an Offering

In many traditions, acknowledging the gift of a dream symbol strengthens your connection to it. You might plant seeds in its honor, create an altar with its image, or simply express gratitude in writing. This reinforces respect for the relationship.

7. Watch for Recurrence

If the same plant appears repeatedly in dreams or visions, take it seriously. Recurrent symbols often indicate a lesson or transformation you are resisting or one that is central to your life path at the moment.

8. Share and Compare Interpretations

Discuss your plant dreams with others, either in spiritual circles, dream groups, or online communities. Hearing how others relate to the same plant can expand your understanding and reveal shared symbolic threads.

Chapter 20: Humans as Partners in Plant Destiny

I still remember the first time I heard the forest speak.

Not through the rustle of leaves or the creak of branches, but through an actual voice, low, resonant, and impossibly patient.

It was the year 2173. After decades of climate collapse, wars over water, and the slow unraveling of ecosystems, humanity had finally stopped pretending we could "manage" nature without listening to it. That's when the Forest Embassy was born as a living council chamber inside a preserved biome where the oldest trees, rare medicinal plants, and humans could meet as equals.

I was there as a junior interpreter. My work wasn't translating human languages. It was translating plant ones. Using a bio-communication interface strapped to my temples, I could feel the electrical pulses from a cedar, the rhythmic sap flows of an oak, and the subtle chemical shifts of medicinal herbs. The technology didn't create words; it created images, sensations, and tones that the human mind learned to interpret like language.

On my first day, I stood before a massive oak with bark so furrowed it looked like the map of a forgotten world. It had been alive since before my great-great-grandparents were born. My interface warmed as its signal merged with my nervous system, and suddenly, my vision blurred into an endless sequence of the seasons: spring buds, summer heat, autumn gold, winter frost. Over and over. The oak's message was simple but profound: We endure by adapting.

The Forest Embassy wasn't just science. It was also a ceremony. Every meeting began with offerings of water, song, and silence drawn from Indigenous traditions passed down through countless generations. Elders explained that reciprocity wasn't a quaint cultural relic. It was survival. Without giving back, there was no relationship. Without a relationship, there was no future.

Over time, certain plants began to "choose" partners. The partnership wasn't random; the plant's electrical signals would spike whenever a specific human entered the biome, and the human would feel an undeniable pull toward it. When it happened to me, it was with a white sage shrub. Its signal wrapped around my heartbeat like a soft echo, and when I reached out to touch its leaves, my interface flared with a flood of warmth and clarity. The elders called this recognition. The scientists called it resonance mapping. I called it love.

Our partnership changed my work. I stopped thinking like a human trying to control nature and started acting like a representative for another intelligence. The sage taught me through sensations, dreams, and subtle cues how to restore damaged soil, how to guide invasive plants into balance, and how to weave its aromatic smoke into rituals that calmed both human nerves and plant stress responses.

In council meetings, I carried its messages not just about preservation, but about renewal. Plants weren't asking us to "save" them; they were asking us to change with them. To grow into something that could live here for centuries without destroying the ground beneath our feet.

Years later, the Forest Embassy became more than a project. It became a model. Cities began planting their own embassy groves, each with human and plant delegates working side by side. And somewhere in that great, slow dialogue between leaf and lung, root and reason, humanity remembered what it had almost forgotten: that we are not separate from the forest. We are one of its many voices.

And if we truly listen, the forest will speak.

In dreams, visions, and ancient stories, plants have always been more than passive scenery. They have been messengers, protectors, teachers, and mirrors of our inner life. These encounters, whether in the quiet of sleep or the pages of sacred tradition, remind us that our relationship with plants has never been one-sided. They reach toward us as much as we reach toward them.

Yet visions and symbolism are only one part of the relationship. In the waking world, there is an equally profound question: if plants can teach, guide, and even choose us, what is our responsibility in return? Beyond admiration or inspiration, how do we enter into a true partnership, one that honors their role in our survival and spiritual life while safeguarding their future alongside ours?

It is here, in the space between reverence and responsibility, that we arrive at the next chapter.

Indigenous Teachings on Reciprocal Stewardship

Across many Indigenous cultures, reciprocity is at the heart of plant-human relationships. The idea is simple yet profound: never take without giving in return. This is not a metaphorical sentiment; it is a daily practice.

Before harvesting medicinal herbs, elders may sprinkle water over the roots, sing to the plant, or offer a pinch of tobacco to the soil. In some traditions, harvesters express their gratitude aloud, acknowledging the plant's sacrifice. These acts are more than ritual; they are agreements between species, ensuring that the cycle of life remains in balance.

Seasonal cycles are also deeply respected. Plants are gathered at the time when their energy is most abundant, and never so much that the population cannot recover. Such mindful timing ensures that the giving and receiving between humans and plants flows like a steady current, year after year.

In many Indigenous worldviews, plants are not seen as "resources" but as relatives. This kinship comes with obligations. Caring for the land, protecting water sources, and ensuring that wild plant communities remain healthy are not just good environmental practices; they are a spiritual covenant.

Stories passed down through generations often speak of plants as members of the community. There are tales of certain trees offering shelter to ancestors in times of danger, or of plants revealing their healing properties in dreams to those who had faithfully cared for them.

Traditional ecological knowledge gathered over centuries through careful observation reminds us that human survival depends on the survival of the ecosystems we live within. Stewardship is therefore not charity; it is mutual survival.

Today, much of modern conservation has focused on damage control, trying to slow the destruction caused by extraction and exploitation. Indigenous stewardship offers a different approach: one rooted in proactive care and partnership.

Instead of taking until scarcity forces a halt, reciprocity teaches us to ensure abundance before taking at all. This mindset shift from "How much can we get?" to "How much can we give so this plant thrives?"—could transform agriculture, forestry, and herbal medicine.

Around the world, community-led conservation projects inspired by Indigenous principles are showing what's possible: forests replanted with native species, wetlands restored, and traditional harvesting methods revived. These efforts are not just protecting biodiversity; they are rebuilding the bond between humans and the living world.

The Belief that Plants "Choose" Humans

In many shamanic and herbalist traditions, the first step in a plant partnership is not choosing; it is being chosen. A healer might notice that a certain plant keeps appearing along their walking path, in their garden, or in their dreams. At first, it may seem like a coincidence, but the pull becomes unmistakable.

Some stories tell of future herbalists dreaming vividly of a plant they have never seen before, only to later encounter it in real life and recognize it instantly. Others describe a strange sense of familiarity, as if the plant were an old friend calling them home.

Such initiations often spark a deep commitment to learning that plant's medicine, spirit, and ecological needs.

The concept of "ally plants" holds that certain plants form ongoing, personal relationships with specific people. These plants act as guides, healers, and even protectors. In ceremonies, an ally plant might be invoked for clarity, courage, or insight.

Herbalists often say that the plant reveals its gifts over time, offering simple remedies first, then deeper teachings to those who work respectfully and consistently. Trust is considered mutual, just as the healer learns to understand the plant's needs, the plant "learns" how best to assist that healer's work.

When humans and plants enter into this kind of partnership, both are changed. The healer's knowledge deepens, but the plant's destiny may also shift. Many plants now spread across continents began as cultivated allies, carried by human hands to new lands where they could flourish. In this view, cultivation is not domination; it is collaboration. Humans provide conditions for the plant to thrive and expand its influence; in return, the plant nourishes, heals, or protects the human community. Over generations, such partnerships can change ecosystems, traditions, and even the trajectory of human history.

Toolkit
Here are simple ways to put this chapter into action in your own life:

1. Begin with Gratitude Rituals
Before harvesting, watering, or repotting, pause to thank the plant. This can be silent or spoken aloud. If appropriate, offer water, a small pinch of herbs, a song, or a simple breath of appreciation. The key is sincerity.
2. Learn the Seasonal Rhythms
Observe when each plant in your care grows, flowers, and rests. Align your harvesting or pruning with its natural cycle. Avoid taking during vulnerable stages like early shoots or late decline unless necessary for the plant's health.
3. Give More Than You Take
For every plant you harvest, consider planting two more. If you wild-harvest, scatter seeds in healthy soil nearby or leave part of the plant untouched so it can regenerate. Think of yourself as part of its survival strategy.
4. Keep a Plant Ally Journal
If a particular plant keeps appearing in your dreams, walks, or daily life, record these encounters. Note the circumstances, your feelings, and any insights. Over time, patterns may emerge that reveal why this plant has "chosen" you.
5. Practice Deep Observation
Spend at least 10 minutes a week simply sitting with your plants—no tools, no phone, no tasks. Notice their scent, shape, color changes, and energy. This strengthens your intuitive connection and builds familiarity.
6. Support Local Plant Communities
Join or volunteer with local seed banks, native plant restoration projects, or community gardens. These actions expand your stewardship beyond your own home and help safeguard plant biodiversity.
7. Create a Reciprocity Plan

Ask yourself: "How can I actively give back to the plant world each season?" This could be by composting, creating pollinator habitats, teaching others about plant care, or protecting wild spaces. Write your plan and review it quarterly.

8. Work with Plants Spiritually

If you feel called, incorporate plants into meditation, prayer, or creative expression. This may be through creating a small altar, brewing a plant-based tea for reflective time, or planting something symbolic of a personal intention.

9. Learn from Indigenous Wisdom (Respectfully)

Seek out books, talks, or community workshops led by Indigenous knowledge keepers. Approach these teachings with humility, understanding that some knowledge is not meant to be taken out of context. Support their communities in return.

10. Recognize Mutual Transformation

Reflect regularly on how caring for plants has changed you and how your care might be shaping the plant's journey. Awareness of this mutual growth reinforces the idea that you are co-creating a shared future.

Chapter 21: Co-Creation and Global Healing

The first time I took part in the Ceremony of a Thousand Seeds, the air felt different, charged, like the moment before a summer storm. It was the year 2125, and for weeks leading up to it, you could feel the anticipation everywhere. Billboards in the cities didn't sell products anymore; they displayed countdowns to planting day, along with quiet reminders: "Plant with care. Plant with love. Plant for all life."

That morning, the world slowed. No traffic roared, no sirens blared, no commerce clattered in the background. Public transport paused mid-route. People stepped out from apartments, from farms, from ocean platforms, and even from lunar bases broadcasting in real time. No matter where you were, you could feel the planet holding its breath.

Every participant received a small, hand-crafted packet in the weeks before the ceremony. The outer wrap was made from pressed plant fiber embossed with a spiraling emblem—a symbol said to represent the infinite bond between humans and nature. Inside were exactly 1,000 seeds, each chosen for the region's unique climate-restoration needs. They were a blend of sturdy native species, medicinal herbs, soil-healing plants, and flowers to draw pollinators. None of them were random. Each seed had a purpose, a story, and a role in the greater whole.

We gathered in small circles: families, friends, neighbors, strangers. At precisely noon Universal Time, the ceremony began with "The Earth Pulse," a sound broadcast across every speaker, screen, and headset on the planet. It was a deep, resonant hum interlaced with gentle crackles and rhythms, recorded from the combined bioelectric signals of millions of plants around the globe. Some said you could hear the forests breathing in it. Others swore they heard an ancient song.

We stood in silence, hands on the seed packets, feeling the pulse reverberate through our bodies. A woman beside me closed her eyes and began to cry. An elder on my other side whispered a prayer in a language I didn't know.

Then, without a single command, we began. Kneeling, pressing, digging, and covering, our movements became a quiet dance of intention. Some were planted in gardens, others along shorelines, some on mountain slopes where drones had carried them. Children drew patterns in the soil before tucking the seeds into place, as though giving each one a blessing. It was not just planting, it was sending a message. We were told to imagine the world we wished these seeds to grow into. I pictured rivers clear enough to drink from, forests teeming with life, and air so clean it carried the scent of flowers for miles. I whispered that vision into my cupped hands before releasing each handful of seeds into the soil. Afterward, we stayed together for hours, not because anyone told us to, but because leaving felt wrong. The air was alive. Across the globe, billions of seeds rested in fresh soil, connected not just by earth and water but by a shared human act. Satellites and drones monitored each planting site, sending gentle reminders to local caretakers in the months ahead. The ceremony was only the beginning; the tending was a year-round promise.

Years later, I visited the same hillside where I had planted my thousand seeds. The ground that once cracked with drought now bloomed with color. Bees moved like living jewels through wildflowers, young trees cast shade, and herbs grew so thick the wind carried their fragrance. It wasn't just my hillside; this transformation had repeated across deserts, city rooftops, tundras, and abandoned industrial sites.

The Ceremony of a Thousand Seeds became more than a tradition. It became our proof that when humanity acts with unity, the Earth responds. Each year's planting wasn't only about what sprouted from the soil; it was about the invisible roots growing between us as people, binding us to the living world we had almost lost but chose together to save.

When the last seeds were pressed into the earth and the final prayers whispered into the wind, the Ceremony of a Thousand Seeds left more than green shoots behind. It awakened something, a sense that each of us had touched a living circuit running through the entire planet. We had not merely planted for ourselves, or even for our communities; we had become part of a vast and conscious network, one in which every root, every leaf, and every human hand played a role.

That day taught me that the Earth is not a backdrop to our lives, but a partner in them. Her forests, meadows, oceans, and deserts are not separate ecosystems but connected threads in a single tapestry of life.

When we plant, tend, and restore, we are not just altering the landscape; we are sending a signal into the deeper fabric of the planet's being. This is the vision of Earth's Consciousness Network: the idea that every plant is a living node in a system far larger than we can see, and that humans are meant to be active participants in its unfolding story.

The deeper we listen to the living world, the more we begin to sense that we are not merely inhabitants of Earth, we are participants in her unfolding consciousness. The plants, animals, waters, and winds are not separate from us; they are threads woven into the same tapestry, and every choice we make sends a ripple across the whole. This chapter explores the vision of a living, conscious Earth, how humans can restore balance through sacred acts of planting and healing, and what the shared future of humans and plants might look like if we choose partnership over exploitation.

The Vision of Earth's Consciousness Network

The Gaia theory proposes that Earth operates as a self-regulating organism, constantly adjusting to maintain the delicate conditions that allow life to flourish. Every forest, wetland, and grassland plays a part in this great balancing act. Plants are not simply background scenery; they are the "green circuitry" of the planet's vitality, producing oxygen, moderating climate, and cycling nutrients through every ecosystem.

Many Indigenous traditions have long held a parallel understanding. They speak of Earth as Mother, Sky as Father, and plants as our elder siblings —beings who carry ancient wisdom and provide for us without demand. This worldview is not poetic fantasy; it is a functional truth that has guided sustainable living for millennia. When science and ancestral knowledge meet, they both point to the same reality: Earth is alive, and we are a part of her body.

Beneath the soil, mycorrhizal fungi link plant roots into vast underground networks, allowing trees to share nutrients, warn of pests, and even support weaker neighbors. This "wood wide web" is more than a curiosity—it is a living model for planetary communication. If forests can share resources and information, perhaps the Earth itself operates as a web of interconnected intelligence.

The interdependence of ecosystems (forests feeding rivers, rivers nourishing wetlands, and wetlands filtering water for grasslands) shows that no habitat is truly isolated. In this view, humans are not meant to be external observers. We are intended to be active nodes within this network, contributing our own form of care, creativity, and protection. Shifting from exploitation to stewardship requires a fundamental change in perspective. Instead of seeing plants as "resources," we can learn to see them as partners in a mutual covenant of care. In spiritual ecology, this role is called Earth guardianship, a conscious commitment to protecting the integrity of the living world.

Human health is inseparable from plant health. The forests that clean our air, the wetlands that filter our water, and the gardens that nourish our bodies are all part of the same life-support system. By tending to them, we tend to ourselves.

Planting Trees and Restoring Ecosystems as Sacred Acts

In many cultures, planting a tree is more than an environmental act; it is a prayer. In some traditions, a tree is planted to mark the birth of a child, ensuring that as the child grows, the tree grows alongside them. Others plant trees at weddings, binding the lives of two people into the life of the Earth. Memorial trees are planted to carry the memory of loved ones into the future, rooted in living soil.

When seeds are sown with intention, whether whispered blessings, silent gratitude, or spoken vows, the act becomes a sacred offering. Such rituals acknowledge that planting is not only about adding greenery to the land but also about infusing a place's spirit with vitality.

Restoring damaged landscapes is more than a scientific endeavor; it is planetary medicine. Wetlands act as kidneys, filtering toxins from water. Prairies stabilize soil and protect against erosion. Forests moderate temperature and store carbon. When we heal these places, we heal the entire planetary body.

Many restoration projects now integrate traditional ecological knowledge with modern science. In some cases, reforestation begins with a ceremony, prayers, dances, or offerings before the first seedling touches the ground. These practices reaffirm the truth that ecological recovery is not just physical work; it is spiritual work.

Medicinal gardens have long served as living libraries of cultural wisdom. By growing, tending, and passing down herbal knowledge, communities preserve not only remedies but also stories, songs, and seasonal teachings.

Reviving endangered plants through traditional cultivation is an act of both ecological and cultural renewal. In some places, youth programs teach plant-based wisdom alongside language preservation, ensuring that the next generation inherits not only the plants but the worldview that sustains them.

The Shared Future of Humans and Plants

Co-creation is more than working alongside nature it is listening, learning, and shaping the future together. Mutual-benefit projects, such as food forests, pollinator corridors, and community herbal gardens, demonstrate that the same effort can serve both human needs and ecological health.

When local communities unite ecological goals with cultural or spiritual ones, they weave resilience into the land and the people. These projects show that collaboration with plants is not an idealistic dream; it is a practical necessity for thriving in the future.

Small acts matter. A single garden can feed a neighborhood. A reforestation project in one valley can restore water cycles downstream. Local herbalists can spark a regional movement to protect endangered plants. The ripple effects of these actions are amplified when connected to global networks of gardeners, reforesters, and healers.

The path to planetary healing begins with personal relationships with the plants we see each day, the soil beneath our feet, and the food we grow or gather.

Imagine cities where every rooftop grows food and medicine, where tree-lined streets filter the air, and where wilderness corridors connect urban centers to the larger landscape. Imagine rural areas thriving through regenerative farming that restores the health of both people and land.

In such a future, plants and humans evolve together not through domination, but through partnership. This vision is not only possible; it is necessary. The survival of Earth's consciousness depends on the health of her green allies and the will of her human stewards to protect them.

Toolkit

Here are simple ways to put this chapter into action in your own life:

1. Practice Plant-Centered Intention Setting

Before planting anything, whether it's a houseplant, garden herb, or a tree, take a moment to set a clear intention. This could be a blessing for the plant's growth, a gesture of gratitude for its life, or a wish for the well-being of future generations. By rooting your actions in meaning, you align your planting with both ecological and spiritual care.

2. Participate in Local Restoration Efforts

Find a reforestation project, wetland cleanup, or native plant restoration event in your area. Even one day of participation helps build ecological resilience and connects you to like-minded stewards. If such projects don't exist locally, consider starting a small initiative, such as a neighborhood pollinator garden or a community compost program.

3. Create a Personal or Community "Medicine Garden"

Grow herbs that are native or culturally significant to your region. Learn their uses, harvest them respectfully, and share them with others. This preserves traditional plant knowledge while providing living examples for education and cultural renewal.

4. Build Relationships with Local Plants

Choose one or two plants that grow in your immediate environment and learn everything you can about them, their seasonal changes, medicinal uses, ecological role, and history. Treat them as partners rather than objects, observing how your relationship deepens over time.

5. Weave Ceremony into Ecological Action

Incorporate simple rituals into your acts of care for the land—sing while planting, offer water before harvesting, or place a natural token in the soil as an offering. Ritual helps reframe environmental work from "task" to "relationship."

6. Support Plant and Soil Health in Everyday Life

Reduce your chemical footprint by avoiding synthetic pesticides and fertilizers. Compost food scraps, plant diverse species, and create wildlife habitats, even in small spaces. The daily health of your soil and plants contributes directly to planetary healing.

7. Connect to Global Plant Networks

Join organizations or online communities dedicated to plant conservation, reforestation, or the exchange of herbal knowledge. Share your experiences and learn from others so your local efforts contribute to the global movement.

8. Visualize and Share a Co-Creation Future

Regularly imagine what a harmonious future between humans and plants would look like in your community. Share your vision in conversations, on social media, or at local gatherings. Inspiration spreads quickly when it is rooted in hope and practical action.

Chapter 22: Practices to Deepen Plant–Human Bonds

By the mid-22nd century, cities had become ecosystems of steel, glass, and circuitry, yet at their hearts pulsed something older, quieter, and infinitely wiser. They were called Resonance Gardens. Built not for spectacle but for communion, these gardens were living sanctuaries where plants and humans met halfway, their connection amplified through subtle technologies that translated the silent languages of chlorophyll, roots, and bioelectric pulse.

Each visitor who entered a Resonance Garden left their shoes, their devices, and their noise at the gate. Inside, the air shimmered faintly with ionized freshness as wearable neural harmonizers (thin circlets placed behind the ear) tuned the human nervous system to the electrical rhythms of the plants. Breathing slowed, thoughts softened, and the constant static of the modern mind dissolved. For the first time, many felt what their ancestors had whispered all along: that the green world was speaking, waiting, alive.

Mira was sixteen when she first stepped into her city's garden, a reluctant participant in her mother's idea of "therapy." For years, Mira had felt out of sync with the world. Crowds overwhelmed her, sudden shifts in weather left her anxious, and at times she swore she could "feel" moods emanating from trees or flowers in ways she couldn't explain. At school, she kept quiet, unwilling to share experiences that might make her sound strange. But the garden welcomed strangeness; it invited it.

Guided by the attendants, Mira found herself drawn to a towering camphor tree, more than four centuries old, its trunk wide enough to swallow three people standing hand in hand. The neural harmonizer hummed faintly as she settled beneath its boughs. For a long time, she simply listened: to the low creak of branches, the filtered sunlight, the subtle hush of the wind. And then, something shifted.

It began as a tug in her chest, like the slow, steady beat of a second heart aligning with her own. She closed her eyes and felt waves of calm radiating from the tree, pulses rising and falling in harmony with her breath. Images formed, not words exactly, but impressions. The taste of rain on leaves. The warmth of the soil after a long winter. The slow, unshakable patience of something that had seen centuries pass.

Mira began visiting daily. At first, she sat quietly, unsure if what she felt was real or imagined. But the more she returned, the more her own rhythms fell into step with the tree's. Her restless anxiety softened. Her dreams grew clearer. She began to sense subtle changes even before they occurred: a shift in the tree's energy before a storm, a quiet anticipation in its sap before spring's first bloom. What she learned in silence she could not have found in books.

One afternoon, the camphor's pulse grew troubled. The harmonizer translated faint spikes into tones that wavered with unease. Mira placed her hands on the trunk and felt the warning rush through her body like a chill. The next morning, a violent storm swept across the city. Thanks to Mira's urging, her family had secured their windows and prepared, avoiding what could have been a dangerous night.

Word spread of the girl who could "hear the tree." Some dismissed it as a coincidence, but others began to wonder if the plants were teaching not just Mira, but humanity itself. As her bond deepened, Mira understood what the Resonance Garden was meant to be: not a museum of plants for humans to admire, but a training ground for co-evolution, a place where people remembered how to listen and plants remembered how to teach.

By the time Mira was grown, she had become one of the first recognized "Resonance Guides," helping others form their own partnerships with ancient trees, wildflowers, and medicinal herbs. What began as personal healing blossomed into something far larger, A movement reminding humanity that survival would not come through domination, but through listening, attunement, and shared rhythm.

And so, the Resonance Gardens were no longer seen as quiet retreats. They were cathedrals of green circuitry, nodes of planetary intelligence, where the voices of the past, present, and future converged in a living dialogue.

Mira often thought back to her first day beneath the camphor tree, when she had felt like an outsider to the world. Now she knew the truth: she had never been outside at all. She had been listening to the wrong signals. The green pulse was always there, waiting.

As the bond between humans and plants unfolds across history, culture, and imagination, it becomes clear that this relationship is not simply metaphorical or symbolic; it is lived, felt, and available in the present moment. Ancient ceremonies, Indigenous teachings, and futuristic visions all point toward the same truth: plants are not only companions in survival, but also partners in consciousness.

Yet knowing this truth and embodying it are different things. To truly step into partnership with the green world requires practice, a willingness to slow down, to listen, and to cultivate new ways of perception. The doorway to this relationship does not demand elaborate tools or distant travels. It begins in the garden outside your home, in a nearby forest, or even in a single potted plant that you pay attention to and respect.

Meditation and Mindfulness in Gardens or Forests

One of the most direct ways to commune with plants is to sit quietly in a forest. Forest meditation begins with guided breathing: inhaling deeply while focusing on the fragrance of pine, damp soil, or wildflowers; exhaling slowly as you release distraction. The rhythm of breath aligns with the living breath of the forest itself.

Walking meditation along a natural trail offers another form of practice. By taking slow, deliberate steps, we bring awareness to each footfall, allowing the sounds of rustling leaves, birdsong, and distant creaks of branches to enter our awareness without judgment.

The key is absorption, not analysis. By letting go of the need to name or categorize what we see, we open ourselves to the living presence of the plants, experiencing them as companions rather than objects.

Gardens can become micro-sanctuaries for meditation. Tending to plants with mindful attention—whether watering, pruning, or planting transforms ordinary acts into spiritual practice. Each gesture can carry the intention of care and reciprocity.

Mindfulness in the garden also involves attuning to the seasonal cycles. Observing the tender shoots of spring, the blossoming of summer, the harvest of autumn, and the dormancy of winter allows us to mirror these rhythms within our own lives. In this way, gardens become teachers of impermanence, patience, and renewal.

Many traditions recognize the sacredness of practicing among plants. Yoga performed in a garden, qigong beneath an ancient tree, or prayer within a forest chapel creates a direct fusion of human spirit and natural vitality.

Spiritual leaders across cultures, whether the Buddha beneath the bodhi tree or Celtic druids in oak groves, have practiced meditation in the company of plants. Drawing inspiration from these examples, each person can create personal rituals: bowing before entering a forest, placing hands on the soil before beginning practice, or lighting incense as an offering. Such gestures acknowledge that we are entering a shared space of life and awareness.

Plant Attunement Exercises

Deep connection often begins with focus. The "One Plant" practice invites you to select a single plant, perhaps a tree in your neighborhood, a flower in your garden, or a potted herb indoors, and visit it daily.

Each day, observe its growth: the unfolding of new leaves, the shift in scent, or the subtle changes in energy. Keep a record of your impressions. Over time, you may notice that your own moods and inner states resonate with the plant's transformations, weaving your life into its cycle.

Plant attunement is not only visual. By sitting close to a plant, you can notice textures of bark or leaf, the warmth or coolness of its surface, the sound of wind moving through its branches, or the creak of its stem in silence.

Smelling soil, leaves, or blossoms activates deep sensory memory, strengthening your bond with the plant. These moments of immersion awaken the subtle layers of perception that modern life often dulls.

Many traditions teach that plants respond to respect. Before touching or harvesting, pause to ask permission in silence. Whether or not words arise, this simple act transforms your relationship from taking to participating.

Offerings of water, song, or gratitude acknowledge the reciprocity of the exchange. Over time, you may develop your own language of gestures, such as placing a hand on the soil, humming a melody, or whispering thanks, as a personal ritual of honoring the plant world.

Journaling with Plant Allies

Journaling bridges outer experience with inner awareness. After time with plants, write down your sensory impressions, moods, and insights. Track how encounters shift your emotional state or open new perspectives. Over seasons and years, your journal becomes a living record of your relationship with the green world.

To deepen the connection, write as though the plant is speaking back. Use prompts such as, "What do you want me to know today?" or "How can I support you?" Often, the responses that arise, whether poetic, symbolic, or surprising, reveal aspects of your own unconscious filtered through the plant's presence. Reviewing past entries may uncover patterns and guidance you had not noticed at first.

Plants inspire creativity. Poetry, sketches, or photography drawn from plant encounters carry the energy of your connection. Some practitioners incorporate pressed leaves or petals into their journals, embedding the plant's physical essence alongside their written reflections.

Over time, these creative expressions may evolve into meditations, stories, or rituals to be shared with others—extending the bond between human and plant into community and culture.

In practicing these methods, we move from theory to relationship, from idea to lived connection. Each encounter with the plant world has the potential to reshape not only our inner lives but also the way we walk upon the Earth.

Toolkit

Here are simple ways to put this chapter into action in your own life:

1. Forest Breathing Practice

Visit a forest or wooded area. Sit quietly, breathe slowly, and focus on the sounds, scents, and shifting light. Let your senses receive without analysis. Practice for 10–15 minutes to align your breath with the forest's rhythm.

2. Garden Mindfulness

Choose a straightforward task: watering, weeding, or planting. Perform it with full attention. Notice textures, smells, and the feel of the soil. Let the act itself be your meditation.

3. Create a Nature Ritual

Before entering a garden or forest, pause. Offer a bow, a breath, or a word of gratitude. Develop a personal gesture that acknowledges plants as living companions.

4. The "One Plant" Practice

Select a single plant to observe daily for at least one month. Record its changes and your impressions in a journal. Notice how its cycle reflects your own inner states.

5. Sensory Immersion

Spend five minutes close to one plant. Feel the texture of its leaves or bark, listen to the sounds around it, and breathe in its scent. Write down or sketch what you discover.

6. Respectful Exchange

Before touching or harvesting, silently ask permission. Afterward, offer something in return water, a song, or gratitude. This transforms your relationship from taking to a partnership.

7. Daily Plant Journal

After time with plants, write a short reflection: what you saw, heard, smelled, or felt. Note any emotions or insights. Over time, this will create a personal map of your plant relationships.

8. Dialogue with a Plant

In your journal, write a conversation with a plant as though it were speaking back. Use prompts like: "What do you want me to know?" or "How can I help you?" Let intuition guide the answers.

9. Creative Expression

Translate your plant encounters into poetry, sketches, or photographs. Consider pressing a leaf or flower into your journal. Share your creations with others to extend the bond beyond yourself.

10. Seasonal Reflection

At the change of each season, return to your "One Plant" or garden. Reflect on what has shifted in the plants, and in yourself. Record insights to deepen awareness of cycles.

This toolkit is meant to be flexible. You don't need to do everything at once. Start with one or two practices that call to you and build gradually. Over time, you'll discover that even small, mindful interactions with plants open the door to a more profound sense of connection, reciprocity, and guidance.

Chapter 23: Building a Plant-Rich Life

By the dawn of the next century, humanity had begun to move in rhythm with the plants once again. The old industrial calendars, based solely on workweeks and fiscal quarters, had proven hollow. They measured time, but not meaning. In their place, a new system emerged, one that bound humanity to the cycles of the Earth itself. It was called the Green Calendar.

The Green Calendar was not the invention of one government or one culture. It was born of the convergence of many movements: Indigenous ceremonies honoring the planting of corn, the solstice rituals of old Europe, the forest festivals of Asia, and the sacred ecological teachings long ignored in the race for progress. Over the decades, these traditions fused into a shared global practice. By 2100, nearly every community on Earth will celebrate its seasons not only through weather but through the life of plants.

Spring was the season of renewal. On the first day of planting, millions of people across the globe knelt to the soil in unison, pressing seeds into the Earth. Families planted together in small urban plots, while farmers planted in great sweeping fields. Even those without gardens were encouraged to drop a seed into public parks or community planters. Entire cities paused their normal rhythm for this act of global sowing, with livestreams connecting children in Tokyo planting rice sprouts, elders in Kenya blessing maize seeds, and families in Brazil placing manioc into fertile earth. Each seed was planted with an intention whispered or sung: words of hope for the year ahead.

Summer was the season of abundance. Mid-year festivals filled plazas, fields, and greenhouses. Communities gathered to share what had grown, whether vegetables, flowers, or medicinal herbs. No one ate alone during these weeks. Harvest tables stretched across entire neighborhoods, laden with food grown by many hands. Rituals were held under blossoming trees, where people reflected on what they had nurtured so far, not just in gardens but in their own lives. Laughter echoed, music filled the streets, and the scent of fresh fruit reminded everyone that the Earth was generous when treated as kin.

Autumn was the season of release. This was when pruning, composting, and returning organic matter to the soil became sacred. Families gathered in small circles to reflect on what needed to be let go, both in their gardens and in their hearts. Just as dried leaves fell back to nourish the ground, people offered old griefs, regrets, and failures into community compost piles, symbolically transformed into the fertile matter of future growth. These ceremonies often ended with a fire where herbs were burned, their fragrant smoke carrying prayers of gratitude into the sky.

Winter was the season of endurance. Evergreens like pine, cedar, and holly were honored as symbols of life that continued through darkness. Families brought sprigs of green into their homes, not merely for decoration, but to remember resilience. Communal gatherings were quieter during this time, filled with story-sharing and candlelight. Dried herbs, teas, and preserved foods became the centerpiece of rituals. It was a season of patience, trust, and faith in the return of spring.

Children raised in the Green Calendar grew up knowing not only their birthdays, but their "plant days." When a child was born, a tree or herb was planted in their name. Each year on that day, they visited their plant, sometimes in a family garden, in a community grove, sometimes in a digital "Green Registry" that allowed families to track seedlings planted in distant forests. Over time, the children grew alongside their trees, seeing in them a reflection of their own journey.

Cities themselves transformed under this way of life. Rooftops bloomed with communal gardens, riversides pulsed with green corridors, and public squares became sanctuaries where seasonal rituals united thousands of people. The Green Calendar became more than a way of marking time; it became a spiritual architecture for humanity's relationship with Earth.

In this future, humans no longer ask, "What is the date?" Instead, they asked, "What is the Earth doing?" The answer was always the same: she was breathing, she was cycling, she was offering. And humanity, at last, had learned to answer back.

The vision of global healing is not just an idea for tomorrow; it begins in the choices each of us makes today. While planetary movements and large-scale reforestation shape the destiny of Earth, it is in the intimacy of our homes, gardens, and daily rituals that the most profound transformation takes root. The Green Calendar reminds us that every season carries its wisdom, but living in harmony with plants requires more than festivals. It asks us to weave the green world into the fabric of our lives, our homes, our routines, our celebrations, and even our inner landscapes.

Chapter 23 turns from the collective dream to the personal practice. Here, we explore how anyone, whether living in a city apartment, a rural farmhouse, or somewhere in between, can build a plant-rich life. This is not only about cultivating beauty or sustenance; it is about creating sacred spaces where reciprocity, mindfulness, and spiritual intention take form.

To live a plant-rich life is to step into partnership with the green world, not just when we visit a forest, but every day, in the places we dwell.

Creating Sacred Plant Spaces at Home

An indoor sanctuary is more than a collection of potted plants; it is an intentional space where life breathes alongside you. Plants chosen for health benefits, such as aloe for cleansing air, lavender for calming energy, or pothos for resilience, do more than decorate; they shift the energy of a room. Their green presence harmonizes the atmosphere, softens edges, and balances light and shadow.

Placement is as important as choice. A plant near a window can filter the morning sun into softer hues, while trailing vines create gentle cascades of movement that mirror flowing water. Arranging plants thoughtfully invites natural circulation of both light and energy, creating harmony between the human dwelling and the plant's vitality.

Complementing this greenery with natural elements such as wood, stone, or water features deepens the sanctuary effect. A small tabletop fountain can echo the voice of rivers. A stone bowl filled with moss becomes a miniature forest. Each element whispers the story of Earth, grounding the home in its natural origin.

For those with outdoor access, be it a balcony, patio, or garden, plant spaces can become intentional retreats for rest and renewal. A balcony of herbs may become a daily altar, while a backyard garden can transform into a living temple. Seating areas nestled among flowers invite reflection, while pathways lined with shrubs or grasses slow the pace of walking, guiding one into stillness.

Privacy can be created through living walls or plant screens, shifting the outdoor space into a cocoon of green. In such spaces, the boundary between human and nature softens. Here, one may sip tea, meditate, or simply sit in silence, surrounded by the subtle intelligence of growing things.

Meaning imbues space with spirit. A plant chosen for personal or ancestral significance carries layers of memory and connection: an olive tree for peace, a lotus for spiritual awakening, or corn for abundance and sustenance. These symbolic plants link personal practice with the wider heritage of humanity.

Altars, art, or statues woven into plant spaces extend their role beyond the physical, reminding us that the sacred is both seen and unseen. Scent plants like rosemary, jasmine, or lavender provide sensory anchors, grounding daily life in emotional and spiritual resonance. When a plant space engages sight, scent, sound, and memory, it becomes a sanctuary where the human soul can breathe.

Seasonal Rituals for Connecting with the Green World

Spring offers a natural time for renewal and planting. Seed planting ceremonies can align with personal intentions: each seed a prayer, each sprout a promise. Simple acts, such as offering compost or water back to the soil, affirm gratitude for Earth's generosity. Outdoor meditation during this season can open the heart to renewal, allowing the energy of the rising sap and budding leaves to awaken our own creativity.

Summer is the time of flourishing and community. Gardens and trees are at their peak, and this season calls for celebration. Harvest feasts, flower-sharing rituals, or mid-year reflections beneath a canopy of green root us in gratitude for abundance. Herbal bundles or wreaths, crafted from summer plants, become tangible symbols of vitality and joy. These can be shared with neighbors, friends, or family as gifts of living medicine.

As the seasons turn, autumn reminds us of cycles of letting go. Ritual pruning, composting, or simply reflecting on what must be released mirrors the wisdom of plants. Winter, though quiet, holds a different kind of vitality. Evergreen arrangements symbolize endurance and protection, while indoor rituals with dried herbs, flowers, or preserved foods keep the bond with nature alive through the dormant months. These practices teach us that even in stillness, life continues to breathe beneath the surface.

Combining Practical Gardening with Spiritual Intention
Every gardening task can become an act of meditation. Planting seeds may symbolize hope, watering can embody care, and weeding can reflect the clearing of inner obstacles. By aligning breath and attention with each movement, feeling the soil between the fingers, hearing the rhythm of water poured, and watching sunlight glisten on leaves, gardening becomes an act of embodied presence.

Traditions such as feng shui remind us that spatial arrangement influences the flow of energy. Positioning plants in ways that support harmony invites balance not only in a space but within the lives of those who dwell there. Some people plant according to moon cycles, aligning with celestial rhythms. Others practice companion planting, where plants that support one another's growth are placed side by side, a quiet reminder of community, cooperation, and interdependence.

The circle of giving must remain unbroken. Composting transforms waste into nourishment, teaching that nothing is ever truly lost. Saving seeds ensures continuity, a legacy passed forward to future generations. Donating plants, herbs, or produce extends the circle of reciprocity into the wider community, reminding us that sacred plant life is not only for ourselves but for the healing of the whole.

To build a plant-rich life is to transform the ordinary into the extraordinary. A pot of basil in the kitchen, a ritual of watering at dawn, a seasonal celebration with neighbors, each act strengthens the invisible threads between humanity and the green world. In such a life, plants are not background, but kin; not objects, but companions; not decoration, but destiny.

Toolkit

Here are simple ways to put this chapter into action in your own life:

This toolkit provides simple, grounded ways to turn the wisdom of this chapter into daily practice. Each exercise is designed to be flexible; adapt it to your space, climate, and lifestyle.

1. Indoor Sanctuary Setup
   - Choose three plants for different qualities: one for air purification (e.g., snake plant), one for calming energy (e.g., lavender or peace lily), and one that personally inspires you.
   - Arrange them where you spend the most time by your desk, near your bed, or beside a favorite chair.
   - Add one natural element (stone, shell, wood carving, or water bowl) to complete the space.

2. Outdoor Sacred Corner
   - Create a reflective spot, no matter how small, a single chair under a tree, a balcony corner with herbs, or a path lined with stones.
   - Spend five minutes there daily, simply observing how light, air, and plants shift through the hours.

3. Seasonal Ritual Starter
   - Spring: Plant one seed (in a pot if you lack outdoor space) and speak an intention for the year.
   - Summer: Share a plant, flower, or harvest item with someone else as a gift of abundance.
   - Autumn: Collect fallen leaves and reflect on what you are ready to release.
   - Winter: Create an evergreen or dried herb arrangement to remind you of endurance.

4. Gardening as Meditation
   - While watering, breathe slowly and intentionally. Inhale as you lift the water, exhale as it pours into the soil.
   - While weeding, silently name one thought, habit, or fear you are ready to release.
   - While planting, affirm what you are inviting into your life.

5. Energy Alignment Practices

- Place one plant where you need balance: near the entrance for welcome, near your workspace for focus, or near your rest space for calm.
- Experiment with moon-phase planting (new moon = beginnings, full moon = nurturing growth, waning moon = releasing).
- Try companion planting even with houseplants: pair species with complementary needs in one pot as a reminder of cooperation.

6. Giving Back Ritual

Conclusion — Returning to the Green Heart

Reflection on Plants as Teachers, Healers, and Partners

Throughout this journey, one truth has become clear: plants sustain us in every possible way. They give us oxygen to breathe, food to nourish us, and medicines to heal our bodies. They regulate the climate, buffer storms, store carbon, and weave the ecosystems that allow human civilization to exist. Beyond these physical contributions, research shows that being around plants improves mental health, reduces stress, sharpens focus, and increases creativity.

Science reveals that plants are not passive background scenery. They are active participants, designers of atmosphere, architects of biodiversity, and co-creators of balance. When we look at the evidence, we see that plants shape the very conditions of human survival.

Yet, the story of plants is not only written in data and graphs. Across cultures and centuries, people have seen plants as guides, healers, and wise beings with their own intelligence. Shamans, monks, herbalists, and mystics have turned to them not only for physical remedies but also for insight, strength, and spiritual companionship.

This beyond-science perspective does not dismiss measurable facts; it expands them. It reminds us that the essence of a tree, a flower, or an herb cannot be captured entirely in chemical formulas. Plants live in the space between what we can measure and what we can feel, offering both tangible sustenance and intangible wisdom.

The invitation before us is to shift our perspective: from seeing plants as resources to be consumed, to partners with whom we share existence. Partnership means reciprocity. It means caring for plants with the same devotion they extend to us. It means building a model of mutual benefit in which human innovation and reverence ensure that plants thrive, and plants, in turn, ensure human survival.

To honor this partnership, we must unite two approaches: the scientific stewardship that ensures ecological resilience, and the spiritual reverence that ensures cultural and personal meaning. Together, they guide us toward harmony.

Encouragement to Approach with Respect, Wonder, and Reciprocity

Respect for plants can be practiced in small, everyday ways: harvesting only what we need, avoiding waste, choosing sustainable sources, and protecting wild spaces. It means honoring Indigenous knowledge, which has long taught the importance of reciprocity with the green world. It also means recognizing that plants are not isolated objects; they are integral threads in the larger web of life.

Wonder is the doorway to relationship. When we pause to truly see a plant, how a leaf spirals, how a seed splits open into life, we are reminded of the miraculous unfolding happening all around us. Wonder requires slowing down, setting aside analysis, and simply being present. Sharing these encounters with family, friends, and community multiplies their impact, spreading a culture of reverence.

Reciprocity asks us not only to receive but also to give back. This can take many forms: planting trees, restoring ecosystems, saving seeds, tending gardens, supporting conservation, and choosing local growers who care for the land. Reciprocity also means teaching the next generation to live with gratitude and balance. Hence, the cycle of giving and receiving never ends.

A Final Vision of Human–Plant Harmony

Imagine a future where cities breathe like forests, with green roofs, vertical gardens, and tree-lined avenues. Clean air flows freely, nutritious foods grow within reach of every community, and plant-based medicines are accessible to all. In this vision, biodiversity flourishes alongside human progress, each supporting the other in resilience and abundance. In such a world, people turn to plants not only for survival but also for guidance, comfort, and joy. Gardens become sanctuaries for meditation, trees become companions in prayer, and herbs become carriers of memory and ritual. Communities find strength in traditions rooted in respect for the natural world, fostering connection not only with plants but also with one another.

The story of humanity and plants is not finished; it is still being written. Co-evolution is our destiny: humans as caretakers, plants as partners, together shaping the planet's future. This requires more than policy or science; it requires culture, spirituality, and daily life infused with care for the green heart of Earth.

The message is simple yet profound: the more we nurture plants, the more they will nurture us in body, mind, and spirit. The path forward is not about dominance, but about harmony. By returning to the green heart, we return to ourselves.

Closing Blessing

May you walk gently upon the Earth,
and let green life walk with you.
May each leaf you see remind you of breath,
And each seed reminds you of possibility.
May your hands learn the rhythm of giving and receiving,
so that what you take is always balanced by what you offer back.
May you never forget that the forests, the fields, the flowers, and the herbs
are not silent; they are speaking,
always teaching, always healing, always waiting for you to listen.
Step into the green heart of the world,
and know that it is also your own.